T0275976

Lecture Notes in Physics

Volume 919

Founding Editors

W. Beiglböck
J. Ehlers
K. Hepp
H. Weidenmüller

Editorial Board

M. Bartelmann, Heidelberg, Germany
B.-G. Englert, Singapore, Singapore
P. Hänggi, Augsburg, Germany
M. Hjorth-Jensen, Oslo, Norway
R.A.L. Jones, Sheffield, UK
M. Lewenstein, Barcelona, Spain
H. von Löhneysen, Karlsruhe, Germany
J.-M. Raimond, Paris, France
A. Rubio, Donostia, San Sebastian, Spain
M. Salmhofer, Heidelberg, Germany
S. Theisen, Potsdam, Germany
D. Vollhardt, Augsburg, Germany
J.D. Wells, Ann Arbor, USA
G.P. Zank, Huntsville, USA

The Lecture Notes in Physics

The series Lecture Notes in Physics (LNP), founded in 1969, reports new developments in physics research and teaching-quickly and informally, but with a high quality and the explicit aim to summarize and communicate current knowledge in an accessible way. Books published in this series are conceived as bridging material between advanced graduate textbooks and the forefront of research and to serve three purposes:

- to be a compact and modern up-to-date source of reference on a well-defined topic
- to serve as an accessible introduction to the field to postgraduate students and nonspecialist researchers from related areas
- to be a source of advanced teaching material for specialized seminars, courses and schools

Both monographs and multi-author volumes will be considered for publication. Edited volumes should, however, consist of a very limited number of contributions only. Proceedings will not be considered for LNP.

Volumes published in LNP are disseminated both in print and in electronic formats, the electronic archive being available at springerlink.com. The series content is indexed, abstracted and referenced by many abstracting and information services, bibliographic networks, subscription agencies, library networks, and consortia.

Proposals should be sent to a member of the Editorial Board, or directly to the managing editor at Springer:

Christian Caron
Springer Heidelberg
Physics Editorial Department I
Tiergartenstrasse 17
69121 Heidelberg/Germany
christian.caron@springer.com

More information about this series at http://www.springer.com/series/5304

János K. Asbóth • László Oroszlány • András Pályi

A Short Course on Topological Insulators

Band Structure and Edge States in One
and Two Dimensions

 Springer

János K. Asbóth
Wigner Research Centre for Physics
Hungarian Academy of Sciences
Budapest, Hungary

László Oroszlány
Department of Physics of Complex Systems
Eötvös Loránd University
Budapest, Hungary

András Pályi
Department of Materials Physics
Eötvös Loránd University
Budapest, Hungary

Department of Physics
Budapest University of Technology
    and Economics
Budapest, Hungary

ISSN 0075-8450   ISSN 1616-6361   (electronic)
Lecture Notes in Physics
ISBN 978-3-319-25605-4   ISBN 978-3-319-25607-8   (eBook)
DOI 10.1007/978-3-319-25607-8

Library of Congress Control Number: 2015960963

Springer Cham Heidelberg New York Dordrecht London
© Springer International Publishing Switzerland 2016
This work is subject to copyright. All rights are reserved by the Publisher, whether the whole or part of
the material is concerned, specifically the rights of translation, reprinting, reuse of illustrations, recitation,
broadcasting, reproduction on microfilms or in any other physical way, and transmission or information
storage and retrieval, electronic adaptation, computer software, or by similar or dissimilar methodology
now known or hereafter developed.
The use of general descriptive names, registered names, trademarks, service marks, etc. in this publication
does not imply, even in the absence of a specific statement, that such names are exempt from the relevant
protective laws and regulations and therefore free for general use.
The publisher, the authors and the editors are safe to assume that the advice and information in this book
are believed to be true and accurate at the date of publication. Neither the publisher nor the authors or
the editors give a warranty, express or implied, with respect to the material contained herein or for any
errors or omissions that may have been made.

Cover designer: eStudio Calamar, Berlin/Figueres

Printed on acid-free paper

Springer International Publishing AG Switzerland is part of Springer Science+Business Media
(www.springer.com)

Preface

These lecture notes provide an introduction to some of the main concepts of topological insulators, a branch of solid state physics that is developing at a fast pace. They are based on a one-semester course for MSc and PhD students at Eötvös University, Budapest, which the authors have been giving since 2012.

Our aim is to provide an understanding of the core topics of topological insulators—edge states, bulk topological invariants, bulk–boundary correspondence—with as simple mathematical tools as possible. We restricted our attention to one- and two-dimensional band insulators. We use noninteracting lattice models of topological insulators and build these up gradually to arrive from the simplest one-dimensional case (the Su–Schrieffer–Heeger model for polyacetylene) to two-dimensional time-reversal invariant topological insulators (the Bernevig–Hughes–Zhang model for HgTe). In each case we introduce the model first, discuss its properties, and then generalize. The prerequisite for the reader is quantum mechanics and not much else: solid state physics background is provided as we go along.

Since this is an introduction, rather than a broad overview, we try to be self-contained and give citations to the current literature only where it is absolutely necessary. For a broad overview, including pointers to the original papers and current topics, we refer the reader to review articles and books in the Introduction.

Supporting material for these lecture notes in form of Jupyter notebooks are available online at https://github.com/topologicalbudapest/topins.

Despite our efforts, the book inevitably contains typos, errors, and less comprehensible explanations. We would appreciate if you could inform us of any of those; please send your comments to `janos.asboth@wigner.mta.hu`.

Acknowledgments We are grateful for enlightening discussions on topological insulators with Anton Akhmerov, Andrea Alberti, Carlo Beenakker, and Alberto Cortijo. We thank the feedback we got from participants at the courses at Eötvös University, especially Vilmos Kocsis. A version of this course was given by one of us (J.K.A.) in the PhD program of the University of Geneva, on invitation by Markus Büttiker, which helped shape the course.

The preparation of the first version of these lecture notes was supported by the grant TÁMOP4.1.2.A/1-11/0064. We acknowledge financial support from the Hungarian Scientific Research Fund (OTKA), Grant Nos. PD100373, K108676, NN109651, and from the Marie Curie program of the European Union, Grant No. CIG-293834. J.K.A. and A.P. were supported by the János Bolyai Scholarship of the Hungarian Academy of Sciences.

Budapest, Hungary   János K. Asbóth
September 2015   László Oroszlány
                                                                              András Pályi

Introduction

The band theory of electric conduction was one of the early victories of quantum mechanics in the 1920s. It gave a simple explanation of how some crystalline materials are electric insulators, even though electrons in them can hop from one atom to the next. In the bulk of a band insulator, the electrons occupy eigenstates that form energy bands. In a band insulator, there are no partially filled bands: completely filled bands are separated by an energy gap from completely empty bands; the gap represents the energy cost of mobilizing electrons. In contrast, materials with partially filled bands are conductors, where there are plane wave states available to transmit electrons across the bulk at arbitrarily low energy. Although we now know of situations where band theory is inadequate (e.g., for Mott insulators), it remains one of the cornerstones of solid state physics.

The discovery of the quantum Hall effect (1980) has shown that the simple division into band insulators and metals is not the end of the story, not even in band theory. In the quantum Hall effect, a strong magnetic field confines the motion of electrons in the bulk, but the same field forces them into delocalized edge states on the surface. A two-dimensional metal in strong magnetic field is thus an insulator in the bulk, but conducts along the surface, via a discrete number of completely open edge state channels (in the language of the Landauer–Büttiker formalism). The number of edge state channels was linked to the Chern number, a topological invariant of the occupied bands.

Over the last 20 years, theoretical progress over artificial systems has shown that the external magnetic field is not necessary for an insulator to have robust conducting edge states: instead, the nontrivial topology of the occupied bands is the crucial ingredient. The name *topological insulator* was coined for such systems, and their study became a blossoming branch of solid state physics. Following the theoretical prediction (Bernevig, Hughes, and Zhang [5]), electronic transport measurements confirmed that a thin layer of HgTe is a topological insulator (König et al. [21]). Since that time, a host of materials have been shown to be three-dimensional topological insulators and thin films and quantum wires shown to be two- and one-dimensional topological insulators [2].

The intense theoretical interest in topological insulators has led to signature results, such as the "the periodic table of topological insulators" [29], which shows that similarly to phase transitions in statistical mechanics, it is the dimensionality and the basic symmetries of an insulator that decide whether it can be a topological insulator or not. Although it was derived by different ways of connecting topological insulators of various dimensions and symmetries (the so-called dimensional reduction schemes), the mathematically rigorous proof of the periodic table is still missing.

The field of topological insulators is very active, with many experimental challenges and open theoretical problems, regarding the effects of electron–electron interaction, extra crystalline symmetries, coupling to the environment, etc.

Literature

To get a quick and broad overview of topological insulators, with citations for relevant research papers, we recommend the review papers [7, 17, 25]. For a more in-depth look, there are already a few textbooks on the subject (by Bernevig and Hughes [4] and by Shen [30] and one edited by Franz and Molenkamp [10]). To see the link between momentum-space topology and physics in a broader context, we direct the reader to a book by Volovik [34].

There are also introductory courses on topological insulators with a broad scope. We recommend the lectures by Charles Kane (the video recording of the version given at Veldhoven is freely available online) and the online EdX course on topology in condensed matter by a group of lecturers, with the corresponding material collected at topocondmat.org.

These Lecture Notes

Our aim with this set of lecture notes is to complement the literature cited above: we wish to provide a close look at some of the core concepts of topological insulators with as simple mathematical tools as possible. Using one- and two-dimensional noninteracting lattice models, we explain what edge states and what bulk topological invariants are, how the two are linked (this is known as the *bulk–boundary correspondence*), and the meaning and impact of some of the fundamental symmetries.

To keep things as simple as possible, throughout the course we use noninteracting models for solid state systems. These are described using single-particle lattice Hamiltonians, with the zero of the energy corresponding to the Fermi energy. We use natural units, with $\hbar = 1$ and length measured by the lattice constant.

Contents

Chapter 1
The Su-Schrieffer-Heeger (SSH) Model

We take a hands-on approach and get to know the basic concepts of topological insulators via a concrete system: the Su-Schrieffer-Heeger (SSH) model describes spinless fermions hopping on a one-dimensional lattice with staggered hopping amplitudes. Using the SSH model, we introduce the concepts of the single-particle Hamiltonian, the difference between bulk and boundary, chiral symmetry, adiabatic equivalence, topological invariants, and bulk–boundary correspondence.

1.1 The SSH Hamiltonian

The Su-Schrieffer-Heeger (SSH) model describes electrons hopping on a chain (one-dimensional lattice), with staggered hopping amplitudes, as shown in Fig. 1.1. The chain consist of N unit cells, each unit cell hosting two sites, one on sublattice A, and one on sublattice B. Interactions between the electrons are neglected, and so the dynamics of each electron is described by a single-particle Hamiltonian, of the form

$$\hat{H} = v \sum_{m=1}^{N} \left(|m, B\rangle \langle m, A| + h.c. \right) + w \sum_{m=1}^{N-1} \left(|m + 1, A\rangle \langle m, B| + h.c. \right). \qquad (1.1)$$

Here $|m, A\rangle$ and $|m, B\rangle$, with $m \in \{1, 2, \ldots, N\}$, denote the state of the chain where the electron is on unit cell m, in the site on sublattice A, respectively, B, and $h.c.$ stands for Hermitian Conjugate (e.g., $h.c.$ of $\alpha |m, B\rangle \langle m', A|$ is $\alpha^* |m', A\rangle \langle m, B|$ for any $\alpha \in \mathbb{C}$).

The spin degree of freedom is completely absent from the SSH model, since no term in the Hamiltonian acts on spin. Thus, the SSH model describes spin-polarized electrons, and when applying the model to a real physical system, e.g.,

© Springer International Publishing Switzerland 2016

J.K. Asbóth et al., *A Short Course on Topological Insulators*, Lecture Notes in Physics 919, DOI 10.1007/978-3-319-25607-8_1

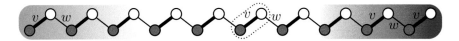

Fig. 1.1 Geometry of the SSH model. Filled (empty) circles are sites on sublattice A (B), each hosting a single state. They are grouped into unit cells: the $n = 6$th cell is circled by a dotted line. Hopping amplitudes are staggered: intracell hopping v (thin lines) is different from intercell hopping w (thick lines). The left and right edge regions are indicated by blue and red shaded background

polyacetylene, we have to always take two copies of it. In this chapter we will just consider a single copy, and call the particles fermions, or electrons, or just particles.

We are interested in the dynamics of fermions in and around the ground state of the SSH model at zero temperature and zero chemical potential, where all negative energy eigenstates of the Hamiltonian are singly occupied (because of the Pauli principle). As we will show later, due to the absence of onsite potential terms, there are N such occupied states. This situation—called half filling—is characteristic of the simplest insulators such as polyacetylene, where each carbon atom brings one conduction electron, and so we find one particle (of each spin type) per unit cell.

For simplicity, we take the hopping amplitudes to be real and nonnegative, $v, w \geq 0$. If this was not the case, if they carried phases, $v = |v| e^{i\phi_v}$, and $w = |w| e^{i\phi_w}$, with $\phi_v, \phi_w \in [0, 2\pi)$ these phases could always be gauged away. This is done by a redefinition of the basis states: $|m, A\rangle \rightarrow e^{-i(m-1)(\phi_v + \phi_w)}$, and $|m, B\rangle \rightarrow e^{-i\phi_v} e^{-i(m-1)(\phi_v + \phi_w)}$.

The matrix for the Hamiltonian of the SSH model, Eq. (1.1), on a real-space basis, for a chain of $N = 4$ unit cells, reads

$$H = \begin{pmatrix} 0 & v & 0 & 0 & 0 & 0 & 0 & 0 \\ v & 0 & w & 0 & 0 & 0 & 0 & 0 \\ 0 & w & 0 & v & 0 & 0 & 0 & 0 \\ 0 & 0 & v & 0 & w & 0 & 0 & 0 \\ 0 & 0 & 0 & w & 0 & v & 0 & 0 \\ 0 & 0 & 0 & 0 & v & 0 & w & 0 \\ 0 & 0 & 0 & 0 & 0 & w & 0 & v \\ 0 & 0 & 0 & 0 & 0 & 0 & v & 0 \end{pmatrix}. \tag{1.2}$$

1.1.1 External and Internal Degrees of Freedom

There is a practical representation of this Hamiltonian, which emphasizes the separation of the external degrees of freedom (unit cell index m) from the internal degrees of freedom (sublattice index α). We can use a tensor product basis,

$$|m, \alpha\rangle \rightarrow |m\rangle \otimes |\alpha\rangle \in \mathcal{H}_{\text{external}} \otimes \mathcal{H}_{\text{internal}}, \tag{1.3}$$

with $m = 1, \ldots, N$, and $\alpha \in \{A, B\}$. In this basis, with the Pauli matrices,

$$\sigma_0 = \begin{pmatrix} 1 & 0 \\ 0 & 1 \end{pmatrix}; \quad \sigma_x = \begin{pmatrix} 0 & 1 \\ 1 & 0 \end{pmatrix}; \quad \sigma_y = \begin{pmatrix} 0 & -i \\ i & 0 \end{pmatrix}; \quad \sigma_z = \begin{pmatrix} 1 & 0 \\ 0 & -1 \end{pmatrix}, \quad (1.4)$$

the Hamiltonian can be written as

$$\hat{H} = v \sum_{m=1}^{N} |m\rangle \langle m| \otimes \hat{\sigma}_x + w \sum_{m=1}^{N-1} \left(|m+1\rangle \langle m| \otimes \frac{\hat{\sigma}_x + i\hat{\sigma}_y}{2} + h.c. \right). \quad (1.5)$$

The intracell hopping shows up here as an intracell operator, while the intercell hopping as a hopping operator.

1.2 Bulk Hamiltonian

As every solid-state system, the long chain of the SSH model has a *bulk* and a *boundary*. The bulk is the long central part of the chain, the boundaries are the two ends, or "edges" of the chain, indicated by shading in Fig. 1.1. We first concentrate on the bulk, since, in the thermodynamic limit of $N \to \infty$, it is much larger than the boundaries, and it will determine the most important physical properties of the model. Although the treatment of the bulk using the Fourier transformation might seem like a routine step, we detail it here because different conventions are used in the literature.

The physics in the bulk, the long central part of the system, should not depend on how the edges are defined, and so for simplicity we set periodic (Born-von Karman) boundary conditions. This corresponds to closing the bulk part of the chain into a ring, with the bulk Hamiltonian defined as

$$\hat{H}_{\mathrm{bulk}} = \sum_{m=1}^{N} \left(v \, |m, B\rangle \langle m, A| + w \, |(m \bmod N) + 1, A\rangle \langle m, B| \right) + h.c.. \quad (1.6)$$

We are looking for eigenstates of this Hamiltonian,

$$\hat{H}_{\mathrm{bulk}} |\Psi_n(k)\rangle = E_n(k) |\Psi_n(k)\rangle, \quad (1.7)$$

with $n \in \{1, \ldots, 2N\}$.

1.2.1 Bulk Momentum-Space Hamiltonian

Due to the translation invariance of the bulk, Bloch's theorem applies, and we look
for the eigenstates in a plane wave form. We introduce the plane wave basis states
only for the external degree of freedom,

$$|k\rangle = \frac{1}{\sqrt{N}} \sum_{m=1}^{N} e^{imk} |m\rangle, \qquad \text{for } k \in \{\delta_k, 2\delta_k, \ldots, N\delta_k\} \quad \text{with } \delta_k = \frac{2\pi}{N},$$

$$(1.8)$$

where the wavenumber k was chosen to take on values from the first Brillouin zone.
The Bloch eigenstates read

$$|\Psi_n(k)\rangle = |k\rangle \otimes |u_n(k)\rangle; \qquad |u_n(k)\rangle = a_n(k) |A\rangle + b_n(k) |B\rangle. \tag{1.9}$$

The vectors $|u_n(k)\rangle \in \mathcal{H}_{\text{internal}}$ are eigenstates of the *bulk momentum-space
Hamiltonian* $\hat{H}(k)$ defined as

$$\hat{H}(k) = \langle k| \hat{H}_{\text{bulk}} |k\rangle = \sum_{\alpha,\beta \in \{A,B\}} \langle k,\alpha| H_{\text{bulk}} |k,\beta\rangle \cdot |\alpha\rangle \langle\beta|; \tag{1.10}$$

$$\hat{H}(k) |u_n(k)\rangle = E_n(k) |u_n(k)\rangle. \tag{1.11}$$

1.2.2 Periodicity in Wavenumber

Although Eq. (1.9) has a lot to do with the continous-variable Bloch theorem,
$\Psi_{n,k}(x) = e^{ikx} u_{n,k}(x)$, this correspondence is not direct. In a discretization of the
continuous-variable Bloch theorem, the internal degree of freedom would play the
role of the coordinate within the unit cell, which is also transformed by the Fourier
transform. Thus, the function $u_{n,k}(x)$ is periodic in real space, $u_{n,k}(x + 1) = u_{n,k}(x)$,
but not periodic in the Brillouin zone, $u_{n,k+2\pi}(x + 1) \neq u_{n,k}(x)$. Our Fourier
transform acts only on the external degree of freedom, and as a result, we have
periodicity in the Brillouin zone,

$$\hat{H}(k + 2\pi) = \hat{H}(k); \qquad |u_n(k + 2\pi)\rangle = |u_n(k)\rangle. \tag{1.12}$$

This convention simplifies the formulas for the topological invariants immensely.
Note, however, that the other convention, the discretization of the Bloch theorem, is
also widely used in the literature.

As an example, take the SSH model on a chain of $N = 4$ unit cells. Then, by inserting Eq. (1.9) into the Schrödinger equation (1.7), the latter translates to the following matrix eigenvalue equation:

$$
\begin{pmatrix}
0 & v & 0 & 0 & 0 & 0 & 0 & w \\
v & 0 & w & 0 & 0 & 0 & 0 & 0 \\
0 & w & 0 & v & 0 & 0 & 0 & 0 \\
0 & 0 & v & 0 & w & 0 & 0 & 0 \\
0 & 0 & 0 & w & 0 & v & 0 & 0 \\
0 & 0 & 0 & 0 & v & 0 & w & 0 \\
0 & 0 & 0 & 0 & 0 & w & 0 & v \\
w & 0 & 0 & 0 & 0 & 0 & v & 0
\end{pmatrix}
\begin{pmatrix}
a(k)e^{ik} \\
b(k)e^{ik} \\
a(k)e^{2ik} \\
b(k)e^{2ik} \\
a(k)e^{3ik} \\
b(k)e^{3ik} \\
a(k)e^{Nik} \\
b(k)e^{Nik}
\end{pmatrix}
= E(k)
\begin{pmatrix}
a(k)e^{ik} \\
b(k)e^{ik} \\
a(k)e^{2ik} \\
b(k)e^{2ik} \\
a(k)e^{3ik} \\
b(k)e^{3ik} \\
a(k)e^{Nik} \\
b(k)e^{Nik}
\end{pmatrix}.
\tag{1.13}
$$

The Schrödinger equation defining the matrix $H(k)$ of the bulk momentum-space Hamiltonian reads

$$
H(k) = \begin{pmatrix} 0 & v + we^{-ik} \\ v + we^{ik} & 0 \end{pmatrix}; \quad H(k)\begin{pmatrix} a(k) \\ b(k) \end{pmatrix} = E(k)\begin{pmatrix} a(k) \\ b(k) \end{pmatrix}.
\tag{1.14}
$$

1.2.3 The Hopping Is Staggered to Open a Gap

The dispersion relation of the bulk can be read off from Eq. (1.14), using the fact that $\hat{H}(k)^2 = E(k)^2\hat{\mathbb{I}}_2$. This gives us

$$
E(k) = \pm \left| v + e^{-ik}w \right| = \pm \sqrt{v^2 + w^2 + 2vw\cos k}
\tag{1.15}
$$

We show this dispersion relation for five choices of the parameters in Fig. 1.2.

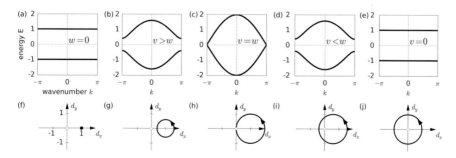

Fig. 1.2 Dispersion relations of the SSH model, Eq. (1.15), for five settings of the hopping amplitudes: (**a**): $v = 1, w = 0$; (**b**): $v = 1, w = 0.6$; (**c**): $v = w = 1$; (**d**): $v = 0.6, w = 1$; (**e**): $v = 0, w = 1$. In each case, the path of the endpoints of the vector $\mathbf{d}(k)$ representing the bulk momentum-space Hamiltonian, Eqs. (1.17) and (1.18), are also shown on the d_x, d_y plane, as the wavenumber is swept across the Brillouin zone, $k = 0 \rightarrow 2\pi$

As long as the hopping amplitudes are staggered, $v \neq w$, (Fig. 1.2a,b,d,e), there is an energy gap of 2Δ separating the lower, filled band, from the upper, empty band, with

$$\Delta = \min_k E(k) = |v - w|. \qquad (1.16)$$

Without the staggering, i.e., if $v = w$, (Fig. 1.2c), the SSH model describes a conductor. In that case there are plane wave eigenstates of the bulk available with arbitrarily small energy, which can transport electrons from one end of the chain to the other.

The staggering of the hopping amplitudes occurs naturally in many solid state systems, e.g., polyacetylene, by what is known as the Peierls instability. A detailed analysis of this process necessitates a model where the positions of the atoms are also dynamical [32]. Nevertheless, we can understand this process intuitively just from the effects of a slight staggering on the dispersion relation. As the gap due to the staggering of the hopping amplitudes opens, the energy of occupied states is lowered, while unoccupied states move to higher energies. Thus, the staggering is energetically favourable.

1.2.4   Information Beyond the Dispersion Relation

Although the dispersion relation is useful to read off a number of physical properties of the bulk of the system (e.g., group velocities), there is also important information about the bulk that it does not reveal. Stationary states do not only have an energy and wavenumber eigenvalue, but also an internal structure, represented by the components of the corresponding vector $|u_n(k)\rangle \in \mathcal{H}_{\text{internal}}$. We now define a compact representation of this information for the SSH model.

The bulk momentum-space Hamiltonian $\hat{H}(k)$ of any two-band model (i.e., a model with 2 internal states per unit cell), reads

$$H(k) = d_0(k)\hat{\sigma}_0 + d_x(k)\hat{\sigma}_x + d_y(k)\hat{\sigma}_y + d_z(k)\hat{\sigma}_z = d_0(k)\hat{\sigma}_0 + \mathbf{d}(k)\hat{\boldsymbol{\sigma}}. \qquad (1.17)$$

For the SSH model, $d_0(k) = 0$, and the real numbers $d_{x,y,z} \in \mathbb{R}$, the components of the k-dependent 3-dimensional vector $\mathbf{d}(k)$, read

$$d_x(k) = v + w\cos k; \qquad d_y(k) = w\sin k; \qquad d_z(k) = 0. \qquad (1.18)$$

The internal structure of the eigenstates with momentum k is given by the direction in which the vector $\mathbf{d}(k)$ of Eq. (1.18) points (the energy is given by the magnitude of $\mathbf{d}(k)$; for details see Sect. 2.5).

As the wavenumber runs through the Brillouin zone, $k = 0 \rightarrow 2\pi$, the path that the endpoint of the vector $\mathbf{d}(k)$ traces out is a closed circle of radius w on the d_x, d_y plane, centered at $(v, 0)$. For more general 2-band insulators, this path need not be a circle, but it needs to be a closed loop due to the periodicity of the bulk momentum-space Hamiltonian, Eq. (1.12), and it needs to avoid the origin, to describe an insulator. The topology of this loop can be characterized by an integer, the *bulk winding number* v. This counts the number of times the loop winds around the origin of the d_x, d_y plane. For example, in Fig. 1.2f,g, we have $v = 0$, in Fig. 1.2i,j, we have $v = 1$, while in Fig. 1.2h, the winding number v is undefined.

1.3 Edge States

Like any material, the SSH Hamiltonian does not only have a bulk part, but also boundaries (which we refer to as *ends* or *edges*). The distinction between bulk and edge is not sharply defined, it describes the behaviour of energy eigenstates in the thermodynamic limit. In the case we consider in these lecture notes, the bulk is translation invariant, and then the we can distinguish edge states and bulk states by their localized/delocalized behaviour in the thermodynamic limit. We will begin with the fully dimerized limits, where the edge regions can be unambiguously defined. We then move away from these limits, and use a practical definition of edge states.

1.3.1 Fully Dimerized Limits

The SSH model becomes particularly simple in the two fully dimerized cases: if the intercell hopping amplitude vanishes and the intracell hopping is set to 1, $v = 1, w = 0$, or vice versa, $v = 0, w = 1$. In both cases the SSH chain falls apart to a sequence of disconnected dimers, as shown in Fig. 1.3.

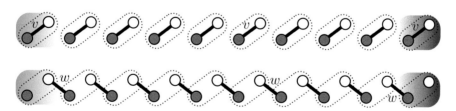

Fig. 1.3 Fully dimerized limits of the SSH model, where the chain has fallen apart to disconnected dimers. In the trivial case (top, only intracell hopping, $v = 1, w = 0$), every energy eigenstate is an even or an odd superposition of two sites at the same unit cell. In the topological case, (bottom, only intercell hopping, $v = 0, w = 1$), dimers are between neighboring unit cells, and there is 1 isolated site per edge, that must contain one zero-energy eigenstate each, as there are no onsite potentials

1.3.1.1   The Bulk in the Fully Dimerized Limits Has Flat Bands

In the fully dimerized limits, one can choose a set of energy eigenstates which are restricted to one dimer each. These consist of the even (energy $E = +1$) and odd (energy $E = -1$) superpositions of the two sites forming a dimer.

In the $v = 1, w = 0$ case, which we call *trivial*, we have

$$v = 1, w = 0: \qquad \hat{H}(|m, A\rangle \pm |m, B\rangle) = \pm(|m, A\rangle \pm |m, B\rangle). \qquad (1.19)$$

The bulk momentum-space Hamiltonian is $\hat{H}(k) = \hat{\sigma}_x$, independent of the wavenumber k.

In the $v = 0, w = 1$ case, which we call *topological*, each dimer is shared between two neighboring unit cells,

$$v = 0, w = 1: \quad \hat{H}(|m, B\rangle \pm |m + 1, A\rangle) = \pm(|m, B\rangle \pm |m + 1, A\rangle), \qquad (1.20)$$

for $m = 1, \ldots, N - 1$. The bulk momentum-space Hamiltonian now is $\hat{H}(k) = \hat{\sigma}_x \cos k + \hat{\sigma}_y \sin k$.

In both fully dimerized limits, the energy eigenvalues are independent of the wavenumber, $E(k) = \pm 1$. In this so-called flat-band limit, the group velocity is zero, which again shows that as the chain falls apart to dimers, a particle input into the bulk will not move along the chain.

1.3.1.2   The Edges in the Fully Dimerized Limit Can Host Zero Energy States

In the trivial case, $v = 1, w = 0$, all energy eigenstates of the SSH chain are given by the formulas of the bulk, Eq. (1.19). A topological, fully dimerized SSH chain, with $v = 0, w = 1$, however, has more energy eigenstates than those listed Eq. (1.20). Each end of the chain hosts a single eigenstate at zero energy,

$$v = 0, w = 1: \qquad \hat{H}|1, A\rangle = \hat{H}|N, B\rangle = 0. \qquad (1.21)$$

These eigenstates have support on one site only. Their energy is zero because onsite potentials are not allowed in the SSH model. These are the simplest examples of *edge states*.

1.3.2   Moving Away from the Fully Dimerized Limit

We now examine what happens to the edge states as we move away from the fully dimerized limit. To be specific, we examine how the spectrum of an open topological

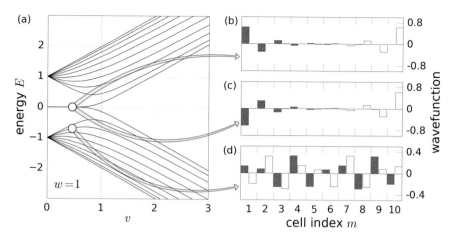

Fig. 1.4 Energy spectrum and wave functions of a finite-sized SSH model. The number of unit cells is $N = 10$. (**a**) Energy spectrum of the system for intercell hopping amplitude $w = 1$ as a function the intracell hopping amplitude v. $v < 1$ ($v > 1$) corresponds to the topological (trivial) phases. (**b**) and (**c**) shows the wave functions of the hybridized edge states, while (**d**) shows a generic bulk wave function

chain, $v = 0, w = 1$, of $N = 10$ unit cells changes, as we continuously turn on the intracell hopping amplitude v. The spectra, Fig. 1.4, reveal that the energies of the edge states remain very close to zero energy.

The wavefunctions of almost-zero-energy edge states have to be exponentially localized at the left/right edge, because the zero of energy is in the bulk band gap. A plot of the wavefunctions (which have only real components, since the Hamiltonian is real), Fig. 1.4, reveals that the almost-zero-energy eigenstates are odd and even superpositions of states localized exponentially on the left and right edge. This is a result of the exponentially small overlap between the left and the right edge states. We will later show, in Sect. 1.5.6, that the edge-state energies are also controlled by this overlap, and are of the order $E = e^{-N/\xi}$, with a localization length $\xi = 1/\log(v/w)$.

There is an important property of the left (right) edge states, which is only revealed by the plot of the wavefunctions, Fig. 1.4. The right edge state has nonvanishing components only on the A sublattice while the left edge state on the B sublattice.

In the following, we show the generality of these properties, and show the link between the bulk winding number and the presence/absence of edge states, known as bulk–boundary correspondence. In the case of the SSH model, all this hinges on a property of the model known as chiral symmetry.

1.4   Chiral Symmetry

In quantum mechanics, we say that a Hamiltonian $\hat{H}$ has a symmetry represented by a unitary operator $\hat{U}$ if

$$\hat{U}\hat{H}\hat{U}^{\dagger} = \hat{H}. \tag{1.22}$$

In case of a symmetry, $\hat{U}$ and $\hat{H}$ can be diagonalized together, and therefore, $\hat{H}$ has no matrix elements between two eigenstates of $\hat{U}$ with different eigenvalues. This can be understood as a superselection rule: if we partition the Hilbert space into different sectors, i.e., eigenspaces of $\hat{U}$, labeled by the corresponding eigenvalues, then the dynamics as defined by $\hat{H}$ can be regarded separately in each sector.

1.4.1   No Unitary Symmetries

A unitary symmetry can be simply made to disappear if we restrict ourselves to one sector of the Hilbert space. This is how we obtained the bulk momentum-space Hamiltonian, in Sect. 1.2, where the symmetry was the lattice translation operator $\hat{U} = |m+1, A\rangle \langle m, A| + |m+1, B\rangle \langle m, B|$, and the labels of the superselection sectors were the quasimomenta k.

1.4.2   A Different Type of Symmetry

The word "symmetry" is also used in a different sense in condensed matter physics. One example is chiral symmetry. We say that a system with Hamiltonian $\hat{H}$ has *chiral symmetry*, if

$$\hat{\Gamma}\hat{H}\hat{\Gamma}^{\dagger} = -\hat{H}, \tag{1.23}$$

with an operator $\hat{\Gamma}$ that is not only unitary, but fulfils some other criteria as well. Notice the extra minus sign on the right hand side. This has important consequences, which we come to later, but first discuss the criteria on the symmetry operator.

First, the chiral symmetry operator has to be unitary and Hermitian, $\hat{\Gamma}^{\dagger} = \hat{\Gamma}$, which can be written succinctly as

$$\hat{\Gamma}^{\dagger}\hat{\Gamma} = \hat{\Gamma}^2 = 1. \tag{1.24}$$

The reason for this requirement is that if the operator $\hat{\Gamma}^2$ was nontrivial, it would represent a unitary symmetry, since

$$\hat{\Gamma}\hat{\Gamma}\hat{H}\hat{\Gamma}\hat{\Gamma} = -\hat{\Gamma}\hat{H}\hat{\Gamma} = \hat{H}. \qquad (1.25)$$

This could still leave room for the chiral symmetry operator to square to a state-independent phase, $\hat{\Gamma}^2 = e^{i\phi}$. However, this can be got rid of by a redefinition of the chiral symmetry, $\Gamma \rightarrow e^{-i\phi/2}\Gamma$.

Second, it is also required that the sublattice operator $\hat{\Gamma}$ be local. The system is assumed to consist of unit cells, and matrix elements of $\hat{\Gamma}$ between sites from different unit cells should vanish. In the SSH chain, this means that for $m \neq m'$, we have $\langle m, \alpha | \hat{\Gamma} | m', \alpha' \rangle = 0$, for any $\alpha, \alpha' \in (A, B)$. To keep things simple, we can demand that the sublattice operator act in the same way in each unit cell (although this is not strictly necessary), its action represented by a unitary operator $\hat{\gamma}$ acting on the internal Hilbert space of one unit cell, i.e.,

$$\hat{\Gamma} = \hat{\gamma} \oplus \hat{\gamma} \oplus \ldots \oplus \hat{\gamma} = \bigoplus_{m=1}^{N} \hat{\gamma}, \qquad (1.26)$$

where N is the number of unit cells.

A third requirement, which is often not explicitly stated, is that the chiral symmetry has to be *robust*. To understand what we mean by that, first note that in solid state physics, we often deal with Hamiltonians with many local parameters that vary in a controlled or uncontrolled way. An example is the SSH model, where the values of the hopping amplitudes could be subject to spatial disorder. We gather all such parameters in a formal vector, and call it $\underline{\xi} \in \Xi$. Here Ξ is the set of all realizations of disorder that we investigate. Instead of talking about the symmetries of a Hamiltonian $\hat{H}$, we should rather refer to symmetries of a set of Hamiltonians $\{\hat{H}(\underline{\xi})\}$, for all $\underline{\xi} \in \Xi$. This set has chiral symmetry represented by $\hat{\Gamma}$ if

$$\forall \underline{\xi} \in \Xi : \qquad \hat{\Gamma}\hat{H}(\underline{\xi})\hat{\Gamma} = -\hat{H}(\underline{\xi}) \qquad (1.27)$$

with the symmetry operator $\hat{\Gamma}$ independent of the parameters $\underline{\xi}$. This is the robustness of the chiral symmetry.

1.4.3 Consequences of Chiral Symmetry for Energy Eigenstates

We now come to the consequences of chiral symmetry, which are very different from those of conventional symmetries, due to the extra minus sign in its definition, Eq. (1.23).

1.4.3.1 Sublattice Symmetry

Chiral symmetry is also called *sublattice symmetry*. Given the chiral symmetry operator $\hat{\Gamma}$, we can define orthogonal sublattice projectors $\hat{P}_A$ and $\hat{P}_B$, as

$$\hat{P}_A = \tfrac{1}{2}\left(\mathbb{I} + \hat{\Gamma}\right); \qquad\qquad \hat{P}_B = \tfrac{1}{2}\left(\mathbb{I} - \hat{\Gamma}\right), \qquad\qquad (1.28)$$

where $\mathbb{I}$ represents the identity operator on the Hilbert space of the system. Note that $\hat{P}_A + \hat{P}_B = \mathbb{I}$, and $\hat{P}_A\hat{P}_B = 0$. The defining relation of sublattice symmetry, Eq. (1.23), can be written in an equivalent form by requiring that the Hamiltonian induces no transitions from any site on one sublattice to any site on the same sublattice,

$$\hat{P}_A\hat{H}\hat{P}_A = P_B\hat{H}\hat{P}_B = 0; \qquad\qquad \hat{H} = \hat{P}_A\hat{H}\hat{P}_B + \hat{P}_B\hat{H}\hat{P}_A. \qquad (1.29)$$

In fact, using the projectors $\hat{P}_A$ and $\hat{P}_B$ is an alternative and equivalent way of defining chiral symmetry.

1.4.3.2 Symmetric Spectrum

The spectrum of a chiral symmetric Hamiltonian is symmetric. For any state with energy E, there is a chiral symmetric partner with energy $-E$. This is simply seen,

$$\hat{H}\,|\psi_n\rangle = E_n\,|\psi_n\rangle \quad \Longrightarrow \quad \hat{H}\hat{\Gamma}\,|\psi_n\rangle = -\hat{\Gamma}\hat{H}\,|\psi_n\rangle = -\hat{\Gamma}E_n\,|\psi_n\rangle = -E_n\hat{\Gamma}\,|\psi_n\rangle\,. \tag{1.30}$$

This carries different implications for nonzero energy eigenstates and zero energy eigenstates.

For $E_n \neq 0$, the states $|\psi_n\rangle$ and $\hat{\Gamma}\,|\psi_n\rangle$ are eigenstates with different energy, and, therefore, have to be orthogonal. This implies that every nonzero energy eigenstate of $\hat{H}$ has equal support on both sublattices,

$$\text{If } E_n \neq 0: \quad 0 = \langle\psi_n|\,\hat{\Gamma}\,|\psi_n\rangle = \langle\psi_n|\,P_A\,|\psi_n\rangle - \langle\psi_n|\,P_B\,|\psi_n\rangle\,. \tag{1.31}$$

For $E_n = 0$, zero energy eigenstates can be chosen to have support on only one sublattice. This is because

$$\text{If } \hat{H}\,|\psi_n\rangle = 0: \quad \hat{H}\hat{P}_{A/B}\,|\psi_n\rangle = \hat{H}\left(|\psi_n\rangle \pm \hat{\Gamma}\,|\psi_n\rangle\right) = 0\,. \tag{1.32}$$

These projected zero-energy eigenstates are eigenstates of $\hat{\Gamma}$, and therefore are chiral symmetric partners of *themselves*.

1.4.4 Sublattice Projectors and Chiral Symmetry of the SSH Model

The Hamiltonian of the SSH model, Eq. (1.1), is *bipartite*: the Hamiltonian includes no transitions between sites with the same sublattice index. The projectors to the sublattices read

$$\hat{P}_A = \sum_{m=1}^{N} |m, A\rangle \langle n, A| \, ; \qquad \hat{P}_B = \sum_{m=1}^{N} |m, B\rangle \langle n, B| \, . \qquad (1.33)$$

Chiral symmetry is represented by the sublattice operator $\hat{\Sigma}_z$, that multiplies all components of a wavefunction on sublattice B by (-1),

$$\hat{\Sigma}_z = \hat{P}_A - \hat{P}_B. \qquad (1.34)$$

Note that this operator has the properties required of the chiral symmetry operator above: it is unitary, Hermitian, and local.

The chiral symmetry of the SSH model is a restatement of the fact that the Hamiltonian is bipartite,

$$\hat{P}_A \hat{H} \hat{P}_A = \hat{P}_B \hat{H} \hat{P}_B = 0; \qquad \Longrightarrow \qquad \hat{\Sigma}_z \hat{H} \hat{\Sigma}_z = -\hat{H}. \qquad (1.35)$$

This relation holds because $\hat{H}$ only contains terms that are multiples of $|m, A\rangle \langle m', B|$, or of $|m, B\rangle \langle m', A|$ with $m, m' \in \mathbb{Z}$. Upon multiplication from the left and the right by $\hat{\Sigma}_z$, such a term picks up a single factor of -1 (because of the multiplication from the left or because of the multiplication from the right). Note that this relation, equivalent to an anticommutation of $\hat{H}$ and $\hat{\Sigma}_z$, holds whether or not the hopping amplitudes depend on position: therefore, the chiral symmetry represented by $\hat{\Sigma}_z$ has the required property of robustness.

1.4.5 Consequence of Chiral Symmetry: Bulk Winding Number for the SSH Model

We now consider the bulk momentum-space Hamiltonian $\hat{H}(k) = \mathbf{d}(k)\hat{\sigma}$.

The path of the endpoint of $\mathbf{d}(k)$, as the wavenumber goes through the Brillouin zone, $k = 0 \to 2\pi$, is a closed path on the d_x, d_y plane. This path has to avoid the origin, $\mathbf{d} = 0$: if there was a k at which $\mathbf{d}(k) = 0$, the gap would close at this k, and we would not be talking about an insulator. Because of chiral symmetry, the vector $\mathbf{d}(k)$ is restricted to lie on the $d_x d_y$ plane,

$$\hat{sigma}_z \hat{H}(k) \hat{sigma}_z = -\hat{H}(k) \qquad \Longrightarrow \qquad d_z(k) = 0. \qquad (1.36)$$

The endpoint of $\mathbf{d}(k)$ is then a closed, directed loop on the plane, and thus has a well defined integer *winding number* about the origin.

1.4.5.1 Winding Number as the Multiplicity of Solutions

The simplest way to obtain the winding number graphically is counting the number of times $\mathbf{d}(k)$ intersects a curve that goes from the origin of the d_x, d_y plane to infinity.

1. Since $\mathbf{d}(k)$ is a directed curve, it has a left side and a right side. Paint the left side blue, the right side red, as shown in Fig. 1.5a.
2. Take a directed curve $\mathscr{L}$ going from 0 to infinity. We can call this the "line of sight to infinity", although it need not be a straight line. A simple choice is the half-infinite line, $d_y = 0$, $d_x \geq 0$. Two other choices are shown in Fig. 1.5.
3. Identify the intersections of $\mathbf{d}(k)$ with $\mathscr{L}$.
4. Each intersection has a signature: this is $+1$ if the line of sight meets it from the blue side, -1 for the red side.
5. The winding number ν is the sum of the signatures.

We now consider how the winding number ν defined above can change under continuous deformations of $\mathscr{L}$ or of $\mathbf{d}(k)$. We only allow for deformations that keep both curves on the plane, maintain $\mathscr{L}$ going from the origin to infinity, and do not create points where $\mathbf{d}(k) = 0$. Due to the deformations the intersections of $\mathscr{L}$ and $\mathbf{d}(k)$ can move, but this does not change ν. They can also appear or disappear, at points where $\mathscr{L}$ and $\mathbf{d}(k)$ touch. However, they can only appear or disappear pairwise: a red and a blue intersection together, which does not change ν. As an example, the two choices of the line of sight $\mathscr{L}$ in Fig. 1.5a, have 1 or 3 intersections, but the winding number is $+1$, for either of them.

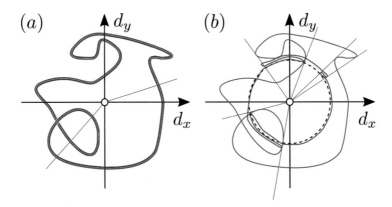

Fig. 1.5 The endpoints of the vector $\mathbf{d}(k)$ as k goes across the Brillouin zone (red or blue closed circles)

1.4.5.2 Winding Number as an Integral

The winding number can also be written as a compact formula using the unit vector $\tilde{\mathbf{d}}$, defined as

$$\tilde{\mathbf{d}} = \frac{\mathbf{d}}{|\mathbf{d}|}. \tag{1.37}$$

This is the result of projecting the curve of $\mathbf{d}(k)$ to the unit circle, as shown in Fig. 1.5b. The vector $\tilde{\mathbf{d}}(k)$ is well defined for all k because $\mathbf{d}(k) \neq 0$.

You can check easily that the winding number ν is given by

$$\nu = \frac{1}{2\pi} \int_{-\pi}^{\pi} \left(\tilde{\mathbf{d}}(k) \times \frac{d}{dk}\tilde{\mathbf{d}}(k) \right)_z dk. \tag{1.38}$$

To calculate ν directly from the bulk momentum-space Hamiltonian, note that it is off-diagonal (in the basis of eigenstates of the chiral symmetry operator $\hat{\sigma}_z$),

$$H(k) = \begin{pmatrix} 0 & h(k) \\ h^*(k) & 0 \end{pmatrix}; \qquad h(k) = d_x(k) - i d_y(k). \tag{1.39}$$

The winding number of $\mathbf{d}(k)$ can be written as an integral, using the complex logarithm function, $\log(|h| e^{i\arg h}) = \log|h| + i\arg h$. It is easy to check that

$$\nu = \frac{1}{2\pi i} \int_{-\pi}^{\pi} dk \frac{d}{dk} \log h(k). \tag{1.40}$$

Here during the calculation of the integral, the branch cut for the logarithm is always shifted so that the derivative is always well defined. The above integral is always real, since $|h(k = -\pi)| = |h(k = \pi)|$.

1.4.5.3 Winding Number of the SSH Model

For the SSH model, the winding number is either 0 or 1, depending on the parameters. In the trivial case, when the intracell hopping dominates the intercell hopping, $v > w$, the winding number is $\nu = 0$. In the topological case, when $w > v$, we have $\nu = 1$.

To change the winding number ν of the SSH model, we need to either (a) pull the path of $\mathbf{d}(k)$ through the origin in the d_x, d_y plane, or (b) lift it out of the plane and put it back on the plane at a different position. This is illustrated in Fig. 1.6. Method (a) requires closing the bulk gap. Method (b) requires breaking chiral symmetry.

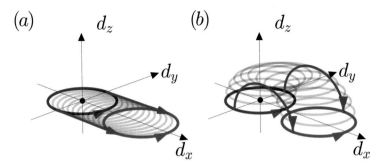

Fig. 1.6 The endpoints of the vector $\mathbf{d}(k)$ as k goes across the Brillouin zone (red or blue closed circles), for various parameter settings in the SSH model. In (**a**), intercell hopping is kept constant at $w = 1$, while the intracell hopping is increased from $v = 0$ to $v = 2.3$. In the process, the bulk gap was closed and reopened, as the origin (black point) falls on one of the blue circles. The winding number is changed from 1 to 0. In (**b**), we again keep $w = 1$, and increase v from 0 to 2.3, but this time avoid closing the bulk gap by introducing a sublattice potential, $H_{\text{sublattice}} = u\hat{\sigma}_z$. We do this by tuning a parameter θ from 0 to π, and setting $v = 1.15(1 - \cos\theta)$, and $u = \sin\theta$. At the end of the process, $\theta = \pi$, there is no sublattice potential, so chiral symmetry is restored. The winding number has been changed from 1 to 0

1.5 Number of Edge States as Topological Invariant

We now introduce the notion of *adiabatic deformation* of insulating Hamiltonians. An insulating Hamiltonian is adiabatically deformed if

- its parameters are changed continuously,
- the important symmetries of the system are maintained,
- the bulk gap around $E = 0$ remains open.

The deformation is a fictitious process, and does not take place in time. However, if we do think of it as a process in real time, the adiabatic theorem [15] tells us, that, starting from the many-body ground state (separated from excited states by the energy gap), and performing the deformation slowly enough, we end up in the ground state, at least as far as the bulk of the system is concerned. At the edges of a system, changes can occur, and there is a subtle point to be made about adiabatic deformations being slow, but not too slow, that the edges should still be considered separately. We will come back to this point in Chap. 4.

1.5.1 Adiabatic Equivalence of Hamiltonians

Two insulating Hamiltonians are said to be *adiabatically equivalent* or *adiabatically connected* if there is an adiabatic deformation connecting them, that respects the important symmetries. For example, in the phase diagram Fig. 1.7 of the SSH model,

Fig. 1.7 Phase diagram of the SSH model. The winding number of the bulk momentum-space Hamiltonian $\hat{H}(k)$ can be $v = 0$, if $v > w$, or $v = 1$, if $v < w$. This defines the trivial (gray) and the topological phase (white). The boundary separating these phases (black solid line), corresponds to $v = w$, where the bulk gap closes at some k. Two Hamiltonians in the same phase are adiabatically connected

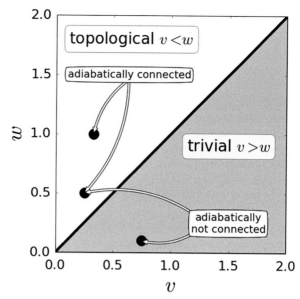

the two Hamiltonians corresponding to the two black points in the topological phase ($w > v$) are adiabatically connected, as one can draw a path between them which does not cross the gapless topological-trivial phase boundary $w = v$.

1.5.2   Topological Invariant

We call an integer number characterizing an insulating Hamiltonian a *topological invariant*, or adiabatic invariant, if it cannot change under adiabatic deformations. Note that the use of adiabatic deformations implies two properties of the topological invariant: (1) it is only well defined in the thermodynamic limit, (2) it depends on the symmetries that need to be respected. An example for a topological invariant is the winding number v of the SSH model.

We know that two insulating Hamiltonians are not adiabatically equivalent if their topological invariants differ. Consider as an example two Hamiltonians corresponding to two points on different sides of the phase boundary in Fig. 1.7 of the SSH model. They are not adiabatically connected in the phase diagram. Nevertheless, one might think that continuously modifying the bulk Hamiltonian by the addition of extra terms, while maintaining chiral symmetry, can lead to a connection between them. However, their winding numbers differ, and since winding numbers cannot change under adiabatic deformation, we know that they are not adiabatically equivalent.

1.5.3   Number of Edge States as a Topological Invariant

We have seen in Sect. 1.3.2, that the number of edge states at one end of the
SSH model was an integer that did not change under a specific type of adiabatic
deformation. We now generalize this example.

Consider energy eigenstates at the left end of a gapped chiral symmetric one-
dimensional Hamiltonian in the thermodynamic limit, i.e., with length $N \to \infty$,
in an energy window from $-\varepsilon < E < \varepsilon$, with ε in the bulk gap. There can be
nonzero-energy edge states in this energy window, and zero-energy edge states as
well. Each nonzero-energy state has to have a chiral symmetric partner, with the
state and its partner occupying the same unit cells (the chiral symmetry operator
is a local unitary). The number of zero-energy states is finite (because of the gap
in the bulk), and they can be restricted to a single sublattice each. There are N_A
zero-energy states on sublattice A, and N_B states on sublattice B.

Consider the effect of an adiabatic deformation of the Hamiltonian, indexed by
some continuous parameter $d : 0 \to 1$, on the number $N_A - N_B$. The Hamiltonian
respects chiral symmetry, and its bulk energy gap exceeds 2ε, for all values of d.

The deformation can create zero-energy states by bringing a nonzero-energy edge
state $|\Psi_0(d = 0)\rangle$ to zero energy, $E_0(d) = 0$ for $d \geq d'$ but not for $d < d'$.
In that case, the chiral symmetric partner of $|\Psi_0\rangle$, which is $\Gamma |\Psi_0(d)\rangle$ up to a
phase factor, has to move simultaneously to zero energy. The newly created zero
energy edge states are $\hat{P}_A |\Psi_0(d')\rangle$ and $\hat{P}_B |\Psi_0(d')\rangle$, which occupy sublattice A and
B, respectively. Thus, the number $N_A - N_B$ is unchanged.

The deformation can also bring a zero energy state $|\Psi_0\rangle$ to energy $E > 0$ at some
$d = d'$. However, it must also create a chiral symmetric partner with energy $E < 0$
at the same d'. This is the time reverse of the process of the previous paragraph:
here, both N_A and N_B must decrease by 1, and, again, $N_A - N_B$ is unchanged.

The deformation can move nonzero-energy states in or out of the $-\varepsilon < E < \varepsilon$
energy window. This obviously has no effect on the number $N_A - N_B$.

Due to the deformation, the wavefunction of a zero-energy eigenstate can change
so that it extends deeper and deeper into the bulk. However, because of the gap
condition, zero-energy states have to have wavefunctions that decay exponentially
towards the bulk, and so this process cannot move them away from the edge. Thus,
N_A and N_B cannot be changed this way.

The arguments above show that $N_A - N_B$, the net number of edge states on
sublattice A at the left edge, is a topological invariant.

1.5.4   Bulk–Boundary Correspondence in the SSH Model

We have introduced two topological invariants for the SSH model: the winding
number ν, of Eq. (1.38), and the net number of edge states, $N_A - N_B$, of this section.
The first one was obtained from the bulk Hamiltonian only, the second by looking at

the low energy sector of the left edge. In the trivial case of the SSH model, $v > w$, both are 0; in the topological case, $v < w$, both are 1. This shows that we can use the bulk topological invariant (the winding number) to make simple robust predictions about the low-energy physics at the edge. This is a simple example for the *bulk–boundary correspondence*, a recurrent theme in the theory of topological insulators, which will reappear in various models in the forthcoming chapters.

1.5.5 Bound States at Domain Walls

Edge states do not only occur at the ends of an open chain, but also at domain walls between different insulating domains of the same chain. This can be understood via the fully dimerized limit. The example in Fig. 1.8 hosts two types of domain walls: one containing a single isolated site, which hosts a zero-energy state (no onsite potentials are allowed), and one containing a trimer. On a trimer, the odd superposition of the two end sites form a zero-energy eigenstate. In the example of Fig. 1.8, this is

$$\hat{H}(|6, B\rangle - |7, B\rangle) = 0. \tag{1.41}$$

Note that, just as the edge states at the ends of the chain, these zero-energy states at the domain walls have wavefunctions that take nonzero values on one sublattice only.

From a perfect dimerized phase without domains it is only possible to germinate an even number of domain walls. This means that if one encounters a domain wall with a localized state on one sublattice then there will be another domain wall somewhere in the system—possibly at the system's edge—with a localized state on the opposite sublattice.

Consider a domain wall in an SSH system that is not in the fully dimerized limit. The wavefunctions of the edge states at the domain walls will penetrate to some small depth into the bulk, with exponentially decaying evanescent tails. For two domain walls at a distance of M unit cells, the two edge states on the walls will hybridize, form "bonding" and "anti-bonding" states. At half filling, of these only the negative energy eigenstate will be occupied. This state hosts a single electron, however, its wavefunction is localized with equal weight on the two domain walls.

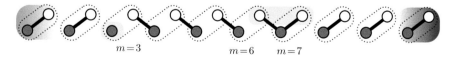

$m=3$ $\qquad\qquad\qquad$ $m=6$ $\qquad$ $m=7$

Fig. 1.8 A long, fully dimerized SSH chain with 3 domains. The boundaries between the domains, the "domain walls", host zero energy eigenstates (yellow shading). These can be localized on a single site, as for the domain wall at $n = 3$, or on a superposition of sites, as the odd superposition of the ends of the trimer shared between the $n = 6$ and $n = 7$ unit cells

Hence each domain wall, when well separated from other domain walls and the ends of the chain, will carry half an electronic charge. This effect is sometimes referred to as "fractionalization" of the charge.

1.5.6 Exact Calculation of Edge States

The zero energy edge states of the SSH model can also be calculated exactly, even in the absence of translational invariance. Take an SSH model on N unit cells, with complex intracell and intercell hopping amplitudes,

$$\hat{H} = \sum_{m=1}^{N} \left(v_m \left| m, B \right\rangle \left\langle m, A \right| + h.c. \right) + \sum_{m=1}^{N-1} \left(w_m \left| m+1, A \right\rangle \left\langle m, B \right| + h.c. \right). \quad (1.42)$$

We are looking for a zero energy eigenstate of this Hamiltonian,

$$\hat{H} \sum_{m=1}^{N} \left(a_m \left| m, A \right\rangle + b_m \left| m, B \right\rangle \right) = 0. \quad (1.43)$$

This gives us $2N$ equations for the amplitudes a_m and b_m, which read

$$m = 1, \ldots, N-1: \qquad v_m a_m + w_m a_{m+1} = 0; \qquad w_m b_m + v_{m+1} b_{m+1} = 0; \quad (1.44a)$$

$$\text{boundaries}: \qquad v_N a_N = 0; \qquad v_1 b_1 = 0. \quad (1.44b)$$

The first set of equations is solved by

$$m = 2, \ldots, N: \qquad a_m = \prod_{j=1}^{m-1} \frac{-v_j}{w_j} a_1; \quad (1.45)$$

$$m = 1, \ldots, N-1: \qquad b_m = \frac{-v_N}{w_m} \prod_{j=m+1}^{N-1} \frac{-v_j}{w_j} b_N. \quad (1.46)$$

However, we also have to fulfil Eq. (1.44b), which give

$$b_1 = a_N = 0. \quad (1.47)$$

These equations together say that, in the generic case, there is no zero energy eigenstate, $a_m = b_m = 0$.

Although there is no exactly zero energy state, Eqs. (1.45), (1.46) and (1.47) admit two approximate solutions in the thermodynamic limit, $N \to \infty$, if the average intercell hopping is stronger than the intracell hopping. More precisely, we define the "bulk average values",

$$\overline{\log |v|} = \frac{1}{N-1} \sum_{m=1}^{N-1} \log |v_m| ; \qquad \overline{\log |w|} = \frac{1}{N-1} \sum_{m=1}^{N-1} \log |w_m| . \qquad (1.48)$$

Equations (1.45) and (1.46) translate to

$$|a_N| = |a_1| \, e^{-(N-1)/\xi} ; \qquad |b_1| = |b_N| \, e^{-(N-1)/\xi} \frac{|v_N|}{|v_1|} , \qquad (1.49)$$

with the localization length

$$\xi = \frac{1}{\overline{\log |w|} - \overline{\log |v|}} . \qquad (1.50)$$

If in the thermodynamic limit, the bulk average values, Eq. (1.48) make sense, and $\xi > 0$, we have two approximate zero energy solutions,

$$|L\rangle = \sum_{m=1}^{N} a_m |m, A\rangle ; \qquad |R\rangle = \sum_{m=1}^{N} b_m |m, B\rangle , \qquad (1.51)$$

with the coefficients a_m and b_m chosen according to Eqs. (1.45) and (1.46), and a_1, respectively, b_N, used to fix the norm of $|L\rangle$, respectively, $|R\rangle$.

1.5.6.1 Hybridization of Edge States

The two states $|L\rangle$ and $|R\rangle$ hybridize under $\hat{H}$ to an exponentially small amount, and this induces a small energy splitting. We can obtain an estimate for the splitting, and the energy eigenstates, to a good approximation using adiabatic elimination of the other eigenstates. In this approximation, the central quantity is the overlap

$$\langle R| \hat{H} |L\rangle = \left| a_1 e^{-(N-1)/\xi} v_N b_N \right| e^{i\phi}, \qquad (1.52)$$

with some $\phi \in [0, 2\pi)$. The energy eigenstates are approximated as

$$|0+\rangle = \frac{e^{-i\phi/2} |L\rangle + e^{i\phi/2} |R\rangle}{\sqrt{2}} ; \qquad E_+ = \left| a_1 e^{-(N-1)/\xi} v_N b_N \right| ; \qquad (1.53)$$

$$|0-\rangle = \frac{e^{-i\phi/2} |L\rangle - e^{i\phi/2} |R\rangle}{\sqrt{2}} ; \qquad E_- = - \left| a_1 e^{-(N-1)/\xi} v_N b_N \right| . \qquad (1.54)$$

The energy of the hybridized states thus is exponentially small in the system size.

Problems

1.1 Higher winding numbers

The SSH model is one-dimensional in space, and has a two-dimensional internal Hilbert space. Construct a lattice model that has these properties of the SSH model, but which has a bulk winding number of 2. Generalize the construction for an arbitrary integer bulk winding number.

1.2 Complex-valued hopping amplitudes

Generalize the SSH model in the following way. Assume that the hopping amplitudes $v = |v|e^{i\phi_v}$ and $w = |w|e^{i\phi_w}$ are complex, and include a third complex-valued hopping amplitude $z = |z|e^{i\phi_z}$ between the states $|m, A\rangle$ and $|m + 1, B\rangle$ for every m. Provide a specific example where the tuning of one of the phases changes the bulk winding number.

1.3 A possible generalization to two dimensions

Consider a two dimensional generalization of the SSH model. Take parallel copies of the SSH chain and couple them without breaking chiral symmetry. What will happen with the edge states?

Chapter 2
Berry Phase, Chern Number

To describe the theory of topological band insulators we will use the language of adiabatic phases. In this chapter we review the basic concepts: the Berry phase, the Berry curvature, and the Chern number. We further describe the relation between the Berry phase and adiabatic dynamics in quantum mechanics. Finally, we illustrate these concepts using a two-level system as a simple example.

For pedagogical introductions, we refer the reader to Berry's original paper [6], and papers from the American Journal of Physics [14, 18]. For the application to solid state physics, we will mostly build on Resta's lecture note [26], and the review paper [36].

2.1 Discrete Case

The subject of adiabatic phases is strongly related to adiabatic quantum dynamics, when a Hamiltonian is slowly changed in time, and the time evolution of the quantum state follows the instantaneous eigenstate of the Hamiltonian. In that context, as time is a continuous variable and the time-dependent Schrödinger equation is a differential equation, the adiabatic phase and the related concepts are expressed using differential operators and integrals. We will arrive to that point later during this chapter; however, we start the discussion using the language of discrete quantum states. Besides the conceptual simplicity, this language also offers an efficient tool for the numerical evaluation of the Chern number, which is an important topological invariant for two-dimensional electron systems.

© Springer International Publishing Switzerland 2016
J.K. Asbóth et al., *A Short Course on Topological Insulators*, Lecture Notes
in Physics 919, DOI 10.1007/978-3-319-25607-8_2

2.1.1 Relative Phase of Two Nonorthogonal Quantum States

In quantum mechanics, the state of a physical system is represented by an equivalence class of vectors in a Hilbert space: a multiplication by a complex phase factor does not change the physical content. A gauge transformation is precisely such a multiplication:

$$|\Psi\rangle \rightarrow e^{i\alpha}\,|\Psi\rangle, \quad \text{with } \alpha \in [0, 2\pi). \tag{2.1}$$

In that sense, the phase of a vector $|\Psi\rangle$ does not represent physical information. We can try to define the relative phase γ_{12} of two nonorthogonal states $|\Psi_1\rangle$ and $|\Psi_2\rangle$ as

$$\gamma_{12} = -\arg\langle\Psi_1 \mid \Psi_2\rangle, \tag{2.2}$$

where $\arg(z)$ denotes the phase of the complex number z, with the specification that $\arg(z) \in (-\pi, \pi]$. Clearly, the relative phase γ_{12} fulfils

$$e^{-i\gamma_{12}} = \frac{\langle\Psi_1 \mid \Psi_2\rangle}{|\langle\Psi_1 \mid \Psi_2\rangle|}. \tag{2.3}$$

However, the relative phase is not invariant under a local gauge transformation,

$$|\Psi_j\rangle \rightarrow e^{i\alpha_j}\,|\Psi_j\rangle \qquad\qquad e^{-i\gamma_{12}} \rightarrow e^{-i\gamma_{12}+i(\alpha_2-\alpha_1)}. \tag{2.4}$$

2.1.2 Berry Phase

Take $N \geq 3$ states in a Hilbert space, order them in a loop, and ask about the phase around the loop. As we show below, the answer—the *Berry phase*—is gauge invariant. For states $|\Psi_j\rangle$, with $j = 1, 2, \ldots, N$, and for the ordered list $L = (1, 2, \ldots, N)$ which define the loop, shown in Fig. 2.1, the Berry phase is defined as

$$\gamma_L = -\arg e^{-i(\gamma_{12}+\gamma_{23}+\ldots+\gamma_{N1})} = -\arg\left(\langle\Psi_1 \mid \Psi_2\rangle \langle\Psi_2 \mid \Psi_3\rangle \ldots \langle\Psi_N \mid \Psi_1\rangle\right). \tag{2.5}$$

To show the gauge invariance of the Berry phase, it can be rewritten as

$$\gamma_L = -\arg \operatorname{Tr}\left(|\Psi_1\rangle\langle\Psi_1| \,|\Psi_2\rangle\langle\Psi_2| \ldots |\Psi_N\rangle\langle\Psi_N|\right). \tag{2.6}$$

Here, we expressed the Berry phase γ_L using projectors that are themselves gauge invariant.

Even though the Berry phase is not the expectation value of some operator, it is a gauge invariant quantity, and as such, it can have a direct physical significance. We will find such a significance, but first, we want to gain more intuition about its behaviour.

2.1.3 Berry Flux

Consider a Hilbert space of quantum states, and a finite two-dimensional square lattice with points labelled by $n, m \in \mathbb{Z}$, $1 \leq n \leq N$, and $1 \leq m \leq M$. Assign a quantum state $|\Psi_{n,m}\rangle$ from the Hilbert space to each lattice site. Say you want to know the Berry phase of the loop L around this set,

$$
\gamma_L = - \arg \exp \left[-i \left(\sum_{n=1}^{N-1} \gamma_{(n,1),(n+1,1)} + \sum_{m=1}^{M-1} \gamma_{(N,m),(N,m+1)} \right. \right.
$$
$$
\left. \left. + \sum_{n=1}^{N-1} \gamma_{(n+1,M),(n,M)} + \sum_{m=1}^{M-1} \gamma_{(1,m+1),(1,m)} \right) \right] \qquad (2.7)
$$

as shown in Fig. 2.1. Although the Berry phase is a gauge invariant quantity, calculating it according to the recipe above involves multiplying together many gauge dependent complex numbers. The alternative route, via Eq. (2.6), involves multiplying gauge independent matrices, and then taking the trace.

There is a way to break the calculation of the Berry phase of the loop down to a product of gauge independent complex numbers. To each plaquette (elementary square) on the grid, with n, m indexing the lower left corner, we define the

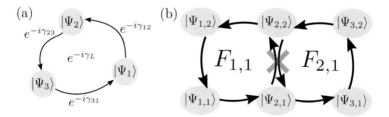

Fig. 2.1 Berry phase, Berry flux and Berry curvature for discrete quantum states. (**a**) The Berry phase γ_L for the loop L consisting of $N = 3$ states is defined from the relative phases $\gamma_{12}, \gamma_{23}, \gamma_{31}$. (**b**) The Berry phase of a loop defined on a lattice of states can be expressed as the sum of the Berry phases $F_{1,1}$ and $F_{2,1}$ of the plaquettes enclosed by the loop. The plaquette Berry phase $F_{n,m}$ is also called Berry flux

Berry flux $F_{n,m}$ of the plaquette using the sum of the relative phases around its boundary,

$$
F_{nm} = -\arg \exp\left[-i\left(\gamma_{(n,m),(n+1,m)} + \gamma_{(n+1,m),(n+1,m+1)}\right.\right.
$$
$$
\left.\left. + \gamma_{(n+1,m+1),(n,m+1)} + \gamma_{(n,m+1),(n,m)}\right)\right], \qquad (2.8)
$$

for $n = 1,\ldots,N$ and $m = 1,\ldots,M$. Note that the Berry flux is itself a Berry phase and is therefore gauge invariant. Alternatively, we can also write

$$
F_{nm} = -\arg\left(\langle \Psi_{n,m} \mid \Psi_{n+1,m}\rangle \langle \Psi_{n+1,m} \mid \Psi_{n+1,m+1}\rangle\right.
$$
$$
\left. \langle \Psi_{n+1,m+1} \mid \Psi_{n,m+1}\rangle \langle \Psi_{n,m+1} \mid \Psi_{n,m}\rangle\right), \qquad (2.9)
$$

Now consider the product of all plaquette phase factors $e^{-iF_{nm}}$,

$$
\prod_{n=1}^{N-1}\prod_{m=1}^{M-1} e^{-iF_{nm}} = \exp\left[-i\sum_{n=1}^{N-1}\sum_{m=1}^{M-1} F_{nm}\right] = \exp\left[-i\sum_{n=1}^{N-1}\sum_{m=1}^{M-1}\left(\gamma_{(n,m),(n+1,m)}\right.\right.
$$
$$
\left.\left. + \gamma_{(n+1,m),(n+1,m+1)} + \gamma_{(n+1,m+1),(n,m+1)} + \gamma_{(n,m+1),(n,m)}\right)\right] \qquad (2.10)
$$

Each internal edge of the lattice is shared between two plaquettes, and therefore occurs twice in the product. However, since we fixed the orientation of the plaquette phases, these two contributions will always be complex conjugates of each other, and cancel each other. Therefore the exponent in the right-hand-side of Eq. (2.10) simplifies to the exponent appearing in Eq. (2.7), implying

$$
\exp\left[-i\sum_{n=1}^{N-1}\sum_{m=1}^{M-1} F_{nm}\right] = e^{-i\gamma_L}. \qquad (2.11)
$$

This result is reminiscent of the Stokes theorem connecting the integral of the curl of a vector field on an open surface and the line integral of the vector field along the boundary of the surface. In Eq. (2.11), the sum of the relative phases, i.e., the Berry phase γ_L, plays the role of the line integral, whereas the double sum of the Berry fluxes plays the role of the surface integral. There is an important difference with respect to the Stokes theorem, namely, the equality of the total Berry flux and the Berry phase is not guaranteed: Eq. (2.11) only tells that they are either equal or have a difference of 2π times an integer.

2.1.4 Chern Number

Consider states in a Hilbert space arranged on a grid as above, $|\Psi_{n,m}\rangle$, with $n, m \in \mathbb{Z}$, $1 \le n \le N$, and $1 \le m \le M$, but now imagine this grid to be on the surface of a torus. We use the same definition for the Berry flux per plaquette as in (2.9), but now with $n \bmod N + 1$ in place of $n + 1$ and $m \bmod M + 1$ in place of $m + 1$.

The product of the Berry flux phase factors of all plaquettes is now 1,

$$\prod_{m=1}^{M} \prod_{n=1}^{N} e^{-iF_{nm}} = 1. \tag{2.12}$$

The same derivation can be applied as for Eq. (2.11) above, but now every edge is an internal edge, and so all contributions to the product cancel.

The Chern number Q associated to our structure is defined via the sum of the Berry fluxes of all the plaquettes forming the closed torus surface:

$$Q = \frac{1}{2\pi} \sum_{nm} F_{nm}. \tag{2.13}$$

The fact that the Chern number Q is defined via the gauge invariant Berry fluxes ensures that Q itself is gauge invariant. Furthermore, taking the arg of Eq. (2.12) proves that the Chern number Q is an integer.

It is worthwhile to look a little deeper into the discrete formula for the Chern number. We can define modified Berry fluxes $\tilde{F}_{nm}$ as

$$\tilde{F}_{nm} = \gamma_{(n,m),(n+1,m)} + \gamma_{(n+1,m),(n+1,m+1)} + \gamma_{(n+1,m+1),(n,m+1)} + \gamma_{(n,m+1),(n,m)}. \tag{2.14}$$

Since each edge is shared between two neighboring plaquettes, the sum of the modified Berry fluxes over all plaquettes vanishes,

$$\sum_{m=1}^{M} \sum_{n=1}^{N} \tilde{F}_{nm} = 0. \tag{2.15}$$

If $-\pi \le \tilde{F}_{nm} < \pi$, then we have $\tilde{F}_{nm} = F_{nm}$. However, $\tilde{F}_{nm}$ can be outside the range $[-\pi, \pi)$: then as the logarithm is taken in Eq. (2.8), F_{nm} is taken back into $[-\pi, \pi)$ by adding a (positive or negative) integer multiple of 2π. In that case, we say the plaquette nm contains a number $Q_{nm} \in \mathbb{Z}$ of vortices, with

$$Q_{nm} = \frac{F_{nm} - \tilde{F}_{nm}}{2\pi} \in \mathbb{Z}. \tag{2.16}$$

We have found a simple picture for the Chern number: *The Chern number Q, that is, the sum of the Berry fluxes of all the plaquettes of a closed surface, is the number of vortices on the surface*,

$$Q = \frac{1}{2\pi} \sum_{nm} F_{nm} = \sum_{nm} Q_{nm} \in \mathbb{Z}. \tag{2.17}$$

Although we proved it here for the special case of a torus, the derivation is easily generalized to all orientable closed surfaces. We focused on the torus, because this construction can be used as a very efficient numerical recipe to discretize and calculate the (continuum) Chern number of a 2-dimensional insulator [12], to be defined in Sect. 2.2.4.

2.2 Continuum Case

We now assume that instead of a discrete set of states, $\{|\Psi_j\rangle\}$, we have a continuum, $|\Psi(\mathbf{R})\rangle$, where the $\mathbf{R}$'s are elements of some D-dimensional parameter space $\mathscr{P}$.

2.2.1 Berry Connection

We take a smooth directed path $\mathscr{C}$, i.e., a *curve* in the parameter space $\mathscr{P}$,

$$\mathscr{C} : [0, 1) \to \mathscr{P}, \quad t \mapsto \mathbf{R}(t). \tag{2.18}$$

We assume that all components of $|\Psi(\mathbf{R})\rangle$ are smooth, at least in an open neighborhood of the curve $\mathscr{C}$. The relative phase between two neighbouring states on the curve $\mathscr{C}$, corresponding to the parameters $\mathbf{R}$ and $\mathbf{R} + d\mathbf{R}$, is

$$e^{-i\Delta\gamma} = \frac{\langle \Psi(\mathbf{R}) \mid \Psi(\mathbf{R} + d\mathbf{R}) \rangle}{|\langle \Psi(\mathbf{R}) \mid \Psi(\mathbf{R} + d\mathbf{R}) \rangle|}; \quad \Delta\gamma = i \langle \Psi(\mathbf{R})| \nabla_{\mathbf{R}} |\Psi(\mathbf{R})\rangle \cdot d\mathbf{R}, \tag{2.19}$$

obtained to first order in $d\mathbf{R} \to 0$. The quantity multiplying $d\mathbf{R}$ on the right-hand side defines the *Berry connection*,

$$\mathbf{A}(\mathbf{R}) = i \langle \Psi(\mathbf{R}) \mid \nabla_{\mathbf{R}}\Psi(\mathbf{R}) \rangle = -\text{Im} \langle \Psi(\mathbf{R}) \mid \nabla_{\mathbf{R}}\Psi(\mathbf{R}) \rangle. \tag{2.20}$$

Here $|\nabla_{\mathbf{R}}\Psi(\mathbf{R})\rangle$ is defined by requiring for every Hilbert space vector $|\Phi\rangle$, that

$$\langle \Phi \mid \nabla_{\mathbf{R}}\Psi(\mathbf{R}) \rangle = \nabla_{\mathbf{R}} \langle \Phi \mid \Psi(\mathbf{R}) \rangle. \tag{2.21}$$

The second equality in Eq. (2.20) follows from the conservation of the norm, $\nabla_{\mathbf{R}} \langle \Psi(\mathbf{R}) \mid \Psi(\mathbf{R}) \rangle = 0$.

We have seen in the discrete case that the relative phase of two states is not gauge invariant; neither is the Berry connection. Under a gauge transformation, it changes as

$$|\Psi(\mathbf{R})\rangle \to e^{i\alpha(\mathbf{R})} |\Psi(\mathbf{R})\rangle : \qquad \mathbf{A}(\mathbf{R}) \to \mathbf{A}(\mathbf{R}) - \nabla_{\mathbf{R}}\alpha(\mathbf{R}). \tag{2.22}$$

2.2.2 Berry Phase

Consider a closed directed curve $\mathscr{C}$ in parameter space. The Berry phase along the curve is defined as

$$\gamma(\mathscr{C}) = -\arg\exp\left[-i\oint_{\mathscr{C}} \mathbf{A} \cdot d\mathbf{R}\right] \tag{2.23}$$

The Berry phase of a closed directed curve is gauge invariant, since it can be interpreted as a limiting case of the discrete Berry phase, via Eqs. (2.20), (2.19), and (2.5), and the latter has been shown to be gauge invariant.

2.2.3 Berry Curvature

As in the discrete case above, we would like to express the gauge invariant Berry phase as a surface integral of a gauge invariant quantity. This quantity is the *Berry curvature*. Similarly to the discrete case, we consider a two-dimensional parameter space, and for simplicity denote the parameters as x and y. We take a simply connected region $\mathscr{F}$ in this two-dimensional parameter space, with the oriented boundary curve of this surface denoted by $\partial\mathscr{F}$, and consider the continuum Berry phase corresponding to the boundary.

2.2.3.1 Smoothness of the Manifold of States

Before relating the Berry phase to the Berry curvature, an important note on the manifold $|\Psi(\mathbf{R})\rangle$ of considered states is in order. From now on, we consider a manifold of states, living in our two-dimensional parameter space, that is smooth, in the sense that the map $\mathbf{R} \mapsto |\Psi(\mathbf{R})\rangle \langle \Psi(\mathbf{R})|$ is smooth. Importantly, this condition does not necessarily imply that that the function $\mathbf{R} \mapsto |\Psi(\mathbf{R})\rangle$, also referred to as a *gauge* describing our manifold, is smooth. (For further discussion and examples, see Sect. 2.5.1.) Nevertheless, even if the gauge $\mathbf{R} \mapsto |\Psi(\mathbf{R})\rangle$ is not smooth in a point $\mathbf{R}_0$ of the parameter space, one can always find an alternative gauge $|\Psi'(\mathbf{R})\rangle$ which is (i) locally smooth, that is, smooth in the point $\mathbf{R}_0$, and (ii) locally generates the same map as $|\Psi(\mathbf{R})\rangle$, that is, for which $|\Psi'(\mathbf{R})\rangle\langle\Psi'(\mathbf{R})| = |\Psi(\mathbf{R})\rangle\langle\Psi(\mathbf{R})|$ in an infinitesimal neighborhood of $\mathbf{R}_0$. Let us formulate an intuitive argument supporting the latter claim using quantum-mechanical perturbation theory. Take the Hamiltonian $\hat{H}(\mathbf{R}) = -|\Psi(\mathbf{R})\rangle\langle\Psi(\mathbf{R})|$, which can be substituted in the infinitesimal neighborhood of $\mathbf{R}_0$ with $\hat{H}(\mathbf{R}_0 + \Delta\mathbf{R}) = \hat{H}(\mathbf{R}_0) + \Delta\mathbf{R}\cdot(\nabla\hat{H})(\mathbf{R}_0)$. According to first-order perturbation theory, the ground state of the latter is given by

$$|\Psi'(\mathbf{R}_0 + \Delta\mathbf{R})\rangle = |\Psi(\mathbf{R}_0)\rangle - \sum_{n=2}^{D}|\Psi_n(\mathbf{R}_0)\rangle\langle\Psi_n(\mathbf{R}_0)|\,\Delta\mathbf{R}\cdot(\nabla\hat{H})(\mathbf{R}_0)\,|\Psi(\mathbf{R}_0)\rangle\,,$$

(2.24)

where the states $|\Psi_n(\mathbf{R}_0)\rangle$ $(n = 2, 3, \ldots, D)$, together with $|\Psi(\mathbf{R}_0)\rangle$, form a basis of the Hilbert space. On the one hand, Eq. (2.24) defines a function that is smooth in $\mathbf{R}_0$, hence the condition (i) above is satisfied. On the other hand, as $|\Psi'(\mathbf{R}_0 + \Delta\mathbf{R})\rangle$ is the ground state of $\hat{H}(\mathbf{R}_0 + \Delta\mathbf{R})$, condition (ii) is also satisfied.

2.2.3.2 Berry Phase and Berry Curvature

Now return to our original goal and try to express the Berry phase as a surface integral of a gauge invariant quantity. We start by relating the Berry phase to its discrete counterpart:

$$\oint_{\partial\mathscr{F}}\mathbf{A}\cdot d\mathbf{R} = \lim_{\Delta x,\Delta y\to 0}\gamma_{\partial\mathscr{F}}\,,$$

(2.25)

where we discretize the parameter space using a square grid of steps Δx, Δy, and express the integral as the discrete Berry phase $\gamma_{\partial\mathscr{F}}$ of a loop approximating $\partial\mathscr{F}$, in the limit of an infinitesimally fine grid. Then, from Eq. (2.25) and the Stokes-type theorem in Eq. (2.11), we obtain

$$\exp\left[-i\oint_{\partial\mathscr{F}}\mathbf{A}\cdot d\mathbf{R}\right] = \lim_{\Delta x,\Delta y\to 0}e^{-i\sum_{nm}F_{nm}}\,,$$

(2.26)

where the nm sum goes for the plaquettes forming the open surface $\mathscr{F}$. Furthermore, let us take a gauge $|\Psi'(\mathbf{R})\rangle$ and the corresponding Berry connection $\mathbf{A}'$ that is smooth in the plaquette nm; this could be $|\Psi(\mathbf{R})\rangle$ and $\mathbf{A}$ if that was already smooth. Then, due to the gauge invariance of the Berry flux we have

$$e^{-iF_{nm}} = e^{-iF'_{nm}}\,,$$

(2.27)

where F'_{nm} is the Berry flux corresponding to the locally smooth gauge. Furthermore, in the limit of an infinitely fine grid it holds that

$$F'_{nm} = A'_x \left(x_n + \frac{\Delta x}{2}, y_m \right) \Delta x + A'_y \left(x_{n+1}, y_m + \frac{\Delta y}{2} \right) \Delta y$$
$$- A'_x \left(x_n + \frac{\Delta x}{2}, y_{m+1} \right) \Delta x - A'_y \left(x_n, y_m + \frac{\Delta y}{2} \right) \Delta y. \tag{2.28}$$

Taylor expansion of the Berry connection around $\mathbf{R}_{nm} = \left(x_n + \frac{\Delta x}{2}, y_n + \frac{\Delta y}{2} \right)$ to first order yields

$$F'_{nm} = \left[\partial_x A'_y(\mathbf{R}_{nm}) - \partial_y A'_x(\mathbf{R}_{nm}) \right] \Delta x \Delta y. \tag{2.29}$$

Thereby, with the definition of the *Berry curvature* as

$$B = \lim_{\Delta x, \Delta y \to 0} \frac{F'_{nm}}{\Delta x \Delta y}, \tag{2.30}$$

we obtain a quantity that is gauge invariant, as it is defined via the gauge invariant Berry flux, and is related to the Berry connection via

$$B = \partial_x A'_y(\mathbf{R}_{nm}) - \partial_y A'_x(\mathbf{R}_{nm}). \tag{2.31}$$

We can rephrase Eq. (2.29) as follows: the Berry flux for the nm plaquette is expressed as the product of the Berry curvature on the plaquette and the surface area of the plaquette.

Substituting Eqs. (2.27) and (2.29) into Eq. (2.26) yields

$$\exp \left[-i \oint_{\partial \mathscr{F}} \mathbf{A} \cdot d\mathbf{R} \right] = \exp \left[-i \int_{\mathscr{F}} B(x, y) dx dy \right], \tag{2.32}$$

which is the continuum version of the result (2.11). Equation (2.32) can also be rephrased as

$$\gamma(\partial \mathscr{F}) = -\arg e^{-i \int_{\mathscr{F}} B(x,y) dx dy}. \tag{2.33}$$

2.2.3.3 A Special Case Where the Usual Stokes Theorem Works

A shortcut towards a stronger result than Eq. (2.32) is offered in the special case when $|\Psi(\mathbf{R})\rangle$ is smooth on the open surface $\mathscr{F}$. Then, a direct application of the two-dimensional Stokes theorem implies

$$\oint_{\partial \mathscr{F}} \mathbf{A} \cdot d\mathbf{R} = \int_{\mathscr{F}} (\partial_x A_y - \partial_y A_x) dx dy = \int_{\mathscr{F}} B dx dy \tag{2.34}$$

Summarizing Eqs. (2.32) and (2.34), we can say that line integral of the Berry connection equals the surface integral of the Berry curvature if the set of states $|\Psi(\mathbf{R})\rangle$ is smooth on $\mathscr{F}$, but they might differ with an integer multiple of 2π otherwise.

2.2.3.4 The Case of the Three-Dimensional Parameter Space

Let us briefly discuss also the case of a three-dimensional parameter space. This will be particularly useful in the context of two-level systems. Starting with the case when the gauge $|\Psi(\mathbf{R})\rangle$ on the two-dimensional open surface $\mathscr{F}$ embedded in the three-dimensional parameter space is smooth in the neighborhood of $\mathscr{F}$, we can directly apply the three-dimensional Stokes theorem to convert the line integral of $\mathbf{A}$ to the surface integral of the curl of $\mathbf{A}$ to obtain

$$\oint_{\partial\mathscr{F}} \mathbf{A} \cdot d\mathbf{R} = \int_{\mathscr{F}} \mathbf{B} \cdot d\mathbf{S}, \tag{2.35}$$

where the Berry curvature is defined as the vector field $\mathbf{B}(\mathbf{R})$ via

$$\mathbf{B}(\mathbf{R}) = \nabla_{\mathbf{R}} \times \mathbf{A}(\mathbf{R}), \tag{2.36}$$

which is gauge invariant as in the two-dimensional case. Even if $|\Psi(\mathbf{R})\rangle$ is not smooth on $\mathscr{F}$, the relation

$$\gamma(\partial\mathscr{F}) = -\arg e^{-i\oint_{\partial\mathscr{F}} \mathbf{A}\cdot d\mathbf{R}} = -\arg e^{-i\int_{\mathscr{F}} \mathbf{B}\cdot d\mathbf{S}} \tag{2.37}$$

holds, similarly to the two-dimensional result Eq. (2.32).

Note furthermore that the Berry phase $\gamma(\partial\mathscr{F})$ for a fixed boundary curve $\partial\mathscr{F}$ is not only gauge invariant, but also invariant against continuous deformations of the two-dimensional surface $\mathscr{F}$ embedded in three dimensions, as long as the Berry curvature is smooth everywhere along the way.

We also remark that although we used the three-dimensional notation here, but the above results can be generalized for any dimensionality of the parameter space.

The notation $\mathbf{A}$ and $\mathbf{B}$ for the Berry connection and Berry curvature suggest that they are much like the vector potential and the magnetic field. This is a useful analogy, for instance, $\nabla_{\mathbf{R}}\mathbf{B} = 0$, from the definition (2.36). Nevertheless, it is not true that in every problem where the Berry curvature is nonzero, there is a physical magnetic field.

2.2.4 Chern Number

In the discrete case, we defined the Chern number as a sum of Berry fluxes for a square lattice living on a torus (or any other orientable closed surface). Here, we take a continuum parameter space that has the topology of a torus. The motivation is that certain physical parameter spaces in fact have this torus topology, and the corresponding Chern number does have physical significance. One example will be the Brillouin zone of a two-dimensional lattice representing a solid crystalline material, where the momentum vectors (k_x, k_y), $(k_x + 2\pi, ky)$, and $(k_x, k_y + 2\pi)$ are equivalent.

Quite naturally, in the continuum definition of the Chern number, the sum of Berry fluxes is replaced by the surface integral of the Berry curvature over the whole of the parameter space $\mathscr{P}$,

$$Q = -\frac{1}{2\pi} \int_{\mathscr{P}} B dx dy. \tag{2.38}$$

As this can be interpreted as a continuum limit of the discrete Chern number, it inherits the properties of the latter: the continuum Chern number is a gauge invariant integer.

For future reference, let us lay down the notation to be used for calculating the Chern numbers of electronic energy bands in two-dimensional crystals. Consider a square lattice for simplicity, which has a square-shaped Brillouin zone as well. Our parameter space $\mathscr{P}$ is the two-dimensional Brillouin zone now, which has a torus topology as discussed above. The parameters are the Cartesian components $k_x, k_y \in [-\pi, \pi)$ of the momentum vector $\mathbf{k}$. The electronic energy bands and the corresponding electron wavefunctions can be obtained from the bulk momentum-space Hamiltonian $\hat{H}(k_x, k_y)$. The latter defines the Schrödinger equation

$$\hat{H}(\mathbf{k}) |u_n(\mathbf{k})\rangle = E_n(\mathbf{k}) |u_n(\mathbf{k})\rangle , \tag{2.39}$$

where $n = 1, 2, \ldots$ is the band index, which has as many possible values as the dimension of the Hilbert space of the internal degree of freedom of our lattice model. Note that defining the Berry connection, the Berry curvature and the Chern number for the nth band is possible only if that band is separated from other bands by energy gaps. The Berry connection of the nth band, in line with the general definition (2.20), reads

$$A_j^{(n)}(\mathbf{k}) = i \langle u_n(\mathbf{k})| \partial_{k_j} |u_n(\mathbf{k})\rangle , \quad \text{for} \quad j = x, y. \tag{2.40}$$

The Chern number of the nth band, in correspondence with Eqs. (2.38) and (2.31), reads

$$Q^{(n)} = -\frac{1}{2\pi} \int_{BZ} dk_x dk_y \left(\frac{\partial A_y^{(n)}}{\partial k_x} - \frac{\partial A_x^{(n)}}{\partial k_y} \right). \tag{2.41}$$

Certain approximations of the band structure theory of electrons provide low-dimensional momentum-space Hamiltonians that can be diagonalized analytically, allowing for an analytical derivation of the Chern numbers of the electronic bands. More often, however, the electronic wave functions are obtained from numerical techniques on a finite-resolution grid of (k_x, k_y) points in the Brillouin zone. In that case, the Chern number of a chosen band can still be effectively evaluated using the discrete version of its definition (2.13).

The Chern number of a band of an insulator is a topological invariant in the following sense. One can imagine that the Hamiltonian describing the electrons on the lattice is deformed adiabatically, that is, continuously and with the energy gaps separating the nth band from the other bands kept open. In this case, the Berry curvature varies continuously, and therefore its integral for the Brillouin zone, which is the Chern number, cannot change as the value of the latter is restricted to integers. If the deformation of the crystal Hamiltonian is such that some energy gaps separating the nth band from a neighboring band is closed and reopened, that is, the deformation of the Hamiltonian is not adiabatic, then the Chern number might change. In this sense, the Chern number is a similar topological invariant for two-dimensional lattice models as the winding number is for the one-dimensional SSH model.

2.3   Berry Phase and Adiabatic Dynamics

In most physical situations of interest, the set of states whose geometric features (Berry phases) we are interested in are eigenstates of some Hamiltonian $\hat{H}$. Take a physical system with D real parameters that are gathered into a formal vector $\mathbf{R} = (R_1, R_2, \ldots, R_D)$. The Hamiltonian is a smooth function $\hat{H}(\mathbf{R})$ of the parameters, at least in the region of interest. We order the eigenstates of the Hamiltonian according to the energies $E_n(\mathbf{R})$,

$$\hat{H}(\mathbf{R}) \, |n(\mathbf{R})\rangle = E_n(\mathbf{R}) \, |n(\mathbf{R})\rangle \,. \tag{2.42}$$

We call the set of eigenstates $|n(\mathbf{R})\rangle$ the *snapshot basis.*

The definition of the snapshot basis involves *gauge fixing*, i.e., specifying the otherwise arbitrary phase prefactor for every $|n(\mathbf{R})\rangle$. This can be a tricky issue: even in cases where a gauge exists where all elements of the snapshot basis are smooth functions of the parameters, this gauge might be very challenging to construct.

We consider the following problem. We assume that the system is initialized with $\mathbf{R} = \mathbf{R}_0$ and in an eigenstate $|n(\mathbf{R}_0)\rangle$ that is in the discrete part of the spectrum, i.e., $E_n(\mathbf{R}) - E_{n-1}(\mathbf{R})$ and $E_{n+1}(\mathbf{R}) - E_n(\mathbf{R})$ are nonzero. At time $t = 0$ we thus have

$$\mathbf{R}(t = 0) = \mathbf{R}_0; \qquad\qquad |\psi(t = 0)\rangle = |n(\mathbf{R}_0)\rangle \,. \tag{2.43}$$

Now assume that during the time $t = 0 \rightarrow T$ the parameter vector $\mathbf{R}$ is slowly changed: $\mathbf{R}$ becomes $\mathbf{R}(t)$, and the values of $\mathbf{R}(t)$ define a continuous directed curve $\mathscr{C}$. Also, assume that $|n(\mathbf{R})\rangle$ is smooth along the curve $\mathscr{C}$. The state of the system evolves according to the time-dependent Schrödinger equation:

$$i\frac{d}{dt}|\psi(t)\rangle = \hat{H}(\mathbf{R}(t))|\psi(t)\rangle. \tag{2.44}$$

Further, assume that $\mathbf{R}$ is varied in such a way that at all times the energy gaps around the state $|n(\mathbf{R}(t))\rangle$ remain finite. We can then choose the rate of variation of $\mathbf{R}(t)$ along the path $\mathscr{C}$ to be slow enough compared to the frequencies corresponding to the energy gap, so the *adiabatic approximation* holds In that case, the system remains in the energy eigenstate $|n(\mathbf{R}(t))\rangle$, only picking up a phase. We are now going to find this phase.

By virtue of the adiabatic approximation, we take as Ansatz

$$|\psi(t)\rangle = e^{i\gamma_n(t)}e^{-i\int_0^t E_n(\mathbf{R}(t'))dt'}|n(\mathbf{R}(t))\rangle. \tag{2.45}$$

For better readability, in the following we often drop the t argument where this leads to no confusion. The time derivative of Eq. (2.45) reads

$$i\frac{d}{dt}|\psi(t)\rangle = e^{i\gamma_n}e^{-i\int_0^t E_n(\mathbf{R}(t'))dt'}\left(-\frac{d\gamma_n}{dt}|n(\mathbf{R})\rangle + E_n(\mathbf{R})|n(\mathbf{R})\rangle + i\left|\frac{d}{dt}n(\mathbf{R})\right\rangle\right). \tag{2.46}$$

To show what we mean by $\left|\frac{d}{dt}n(\mathbf{R}(t))\right\rangle$, we write it out explicitly in terms of a fixed basis, that of the eigenstates at $\mathbf{R} = \mathbf{R}_0$:

$$|n(\mathbf{R})\rangle = \sum_m c_m(\mathbf{R})|m(\mathbf{R}_0)\rangle; \tag{2.47}$$

$$\left|\frac{d}{dt}n(\mathbf{R}(t))\right\rangle = \frac{d\mathbf{R}}{dt}\cdot|\nabla_\mathbf{R}n(\mathbf{R})\rangle = \frac{d\mathbf{R}}{dt}\sum_m \nabla_\mathbf{R}c_m(\mathbf{R})|m(\mathbf{R}_0)\rangle. \tag{2.48}$$

We insert the Ansatz (2.45) into the right hand side of the Schrödinger equation (2.44), use the snapshot eigenvalue relation (2.42), simplify and reorder the Schrödinger equation, and obtain

$$-\frac{d\gamma_n}{dt}|n(\mathbf{R})\rangle + i\left|\frac{d}{dt}n(\mathbf{R})\right\rangle = 0. \tag{2.49}$$

Multiplying from the left by $\langle n(\mathbf{R})|$, and using Eq. (2.48), we obtain

$$\frac{d}{dt}\gamma_n(t) = i\left\langle n(\mathbf{R}(t))\left|\frac{d}{dt}n(\mathbf{R}(t))\right\rangle\right. = \frac{d\mathbf{R}}{dt}i\langle n(\mathbf{R})|\nabla_\mathbf{R}n(\mathbf{R})\rangle. \tag{2.50}$$

We have found that for the directed curve $\mathscr{C}$ in parameter space, traced out by $\mathbf{R}(t)$, there is an adiabatic phase $\gamma_n(\mathscr{C})$, which reads

$$\gamma_n(\mathscr{C}) = \int_{\mathscr{C}} i \langle n(\mathbf{R}) \mid \nabla_{\mathbf{R}} n(\mathbf{R}) \rangle \, d\mathbf{R}. \tag{2.51}$$

A related result is obtained after a similar derivation, if the parameter space of the $\mathbf{R}$ points is omitted and the snapshot basis $|n(t)\rangle$ is parametrized directly by the time variable. Then, the adiabatic phase is

$$\gamma_n(t) = \int_0^t i \langle n(t') \mid \partial_{t'} n(t') \rangle \, dt'. \tag{2.52}$$

Equation (2.51) allows us to formulate the key message of this section as the following. Consider the case of an adiabatic and *cyclic* change of the Hamiltonian, that is, when the curve $\mathscr{C}$ is closed, implying $\mathbf{R}(T) = \mathbf{R}_0$. In this case, the adiabatic phase reads

$$\gamma_n(\mathscr{C}) = \oint_{\mathscr{C}} i \langle n(\mathbf{R}) \mid \nabla_{\mathbf{R}} n(\mathbf{R}) \rangle \, d\mathbf{R}. \tag{2.53}$$

Therefore, the adiabatic phase picked up by the state during a cyclic adiabatic change of the Hamiltonian is equivalent to the Berry phase corresponding to the closed oriented curve representing the Hamiltonian's path in the parameter space.

Two further remarks are in order. First, on the face of it, our derivation seems to do too much. It seems that we have produced an exact solution of the Schrödinger equation. Where did we use the adiabatic approximation? In fact, Eq. (2.50) does not imply Eq. (2.49). For the more complete derivation, showing how the nonadiabatic terms appear, see [15].

The second remark concerns the measurability of the Berry phase. The usual way to experimentally detect phases is by an interferometric setup. This means coherently splitting the wavefunction of the system into two parts, taking them through two adiabatic trips in parameter space, via $\mathbf{R}(t)$ and $\mathbf{R}'(t)$, and bringing the parts back together. The interference only comes from the overlap between the states: it is maximal if $|n(\mathbf{R}(T))\rangle = |n(\mathbf{R}'(T))\rangle$, which is typically ensured if $\mathbf{R}(T) = \mathbf{R}'(T)$. The difference in the adiabatic phases γ_n and γ_n' is the adiabatic phase associated with the closed loop $\mathscr{C}$, which is the path obtained by going forward along $t = 0 \rightarrow T : \mathbf{R}(t)$, then coming back along $t = T \rightarrow 0 : \mathbf{R}'(t)$.

2.4 Berry's Formulas for the Berry Curvature

Berry provided [6] two practical formulas for the Berry curvature. Here we present them in a form corresponding to a three-dimensional parameter space. To obtain the two-dimensional case, where the Berry curvature B is a scalar, one can identify the latter with the component B_z of the three-dimensional case treated below; for generalization to higher than 3 dimensions, see the discussion in Berry's paper [6]. First,

$$B_j = -\text{Im } \epsilon_{jkl} \, \partial_k \, \langle n \mid \partial_l n \rangle = -\text{Im } \epsilon_{jkl} \, \langle \partial_k n \mid \partial_l n \rangle + 0, \tag{2.54}$$

where the second term is 0 because $\partial_k \partial_l = \partial_l \partial_k$ but $\epsilon_{jkl} = -\epsilon_{jlk}$.

To obtain Berry's second formula, inserting a resolution of identity in the snapshot basis in the above equation, we obtain

$$\mathbf{B}^{(n)} = -\text{Im} \sum_{n' \neq n} \langle \nabla n \mid n' \rangle \times \langle n' \mid \nabla n \rangle, \tag{2.55}$$

where the parameter set $\mathbf{R}$ is suppressed for brevity. The term with $n' = n$ is omitted from the sum, as it is zero, since because of the conservation of the norm, $\langle \nabla n \mid n \rangle = -\langle n \mid \nabla n \rangle$. To calculate $\langle n' \mid \nabla n \rangle$, start from the definition of the eigenstate $|n\rangle$, act on both sides with ∇, and then project unto $|n'\rangle$:

$$\hat{H} |n\rangle = E_n |n\rangle; \tag{2.56}$$

$$(\nabla \hat{H}) |n\rangle + \hat{H} |\nabla n\rangle = (\nabla E_n) |n\rangle + E_n |\nabla n\rangle; \tag{2.57}$$

$$\langle n' \mid \nabla \hat{H} \mid n \rangle + \langle n' \mid \hat{H} \mid \nabla n \rangle = 0 + E_n \langle n' \mid \nabla n \rangle. \tag{2.58}$$

Act with $\hat{H}$ towards the left in Eq. (2.58), rearrange, substitute into (2.55), and you obtain the second form of the Berry curvature, which is manifestly gauge invariant:

$$\mathbf{B}^{(n)} = -\text{Im} \sum_{n' \neq n} \frac{\langle n| \nabla \hat{H} |n'\rangle \times \langle n'| \nabla \hat{H} |n\rangle}{(E_n - E_{n'})^2}. \tag{2.59}$$

This shows that the monopole sources of the Berry curvature, if they exist, are the points of degeneracy.

A direct consequence of Eq. (2.59), is that the sum of the Berry curvatures of all eigenstates of a Hamiltonian is zero. If all the spectrum of $\hat{H}(\mathbf{R})$ is discrete along a closed curve $\mathscr{C}$, then one can add up the Berry phases of all the energy eigenstates.

$$\sum_n \mathbf{B}^{(n)} = -\text{Im} \sum_n \sum_{n' \neq n} \frac{\langle n | \nabla_{\mathbf{R}} \hat{H} | n' \rangle \times \langle n' | \nabla_{\mathbf{R}} \hat{H} | n \rangle}{(E_n - E_{n'})^2}$$

$$= -\text{Im} \sum_n \sum_{n' < n} \frac{1}{(E_n - E_{n'})^2} \Big(\langle n | \nabla_{\mathbf{R}} \hat{H} | n' \rangle \times \langle n' | \nabla_{\mathbf{R}} \hat{H} | n \rangle$$

$$+ \langle n' | \nabla_{\mathbf{R}} \hat{H} | n \rangle \times \langle n | \nabla_{\mathbf{R}} \hat{H} | n' \rangle \Big) = 0. \tag{2.60}$$

The last equation holds because $\mathbf{a} \times \mathbf{b} = -\mathbf{b} \times \mathbf{a}$ for any two vectors $\mathbf{a}, \mathbf{b}$.

2.5   Example: The Two-Level System

So far, most of the discussion on the Berry phase and the related concepts have been kept rather general. In this section, we illustrate these concepts via the simplest nontrivial example, that is, the two-level system.

2.5.1   No Continuous Global Gauge

Consider a Hamiltonian describing a two-level system:

$$\hat{H}(\mathbf{d}) = d_x \hat{\sigma}_x + d_y \hat{\sigma}_y + d_z \hat{\sigma}_z = \mathbf{d} \cdot \hat{\boldsymbol{\sigma}}, \tag{2.61}$$

with $\mathbf{d} = (d_x, d_y, d_z) \in \mathbb{R}^3 \backslash \{0\}$. Here, the vector $\mathbf{d}$ plays the role of the parameter $\mathbf{R}$ in of preceding sections, and the parameter space is the punctured three-dimensional Euclidean space $\mathbb{R}^3 \backslash \{0\}$, to avoid the degenerate case of the energy spectrum. Note the absence of a term proportional to σ_0: this would play no role in adiabatic phases. Because of the anticommutation relations of the Pauli matrices, the Hamiltonian above squares to a multiple of the identity operator, $\hat{H}(\mathbf{d})^2 = \mathbf{d}^2 \sigma_0$. Thus, the eigenvalues of $\hat{H}(\mathbf{d})$ have to have absolute value $|\mathbf{d}|$.

A practical graphical representation of $\hat{H}(\mathbf{d})$ is the Bloch sphere, shown in Fig. 2.2. The spherical angles $\theta \in [0, \pi)$ and $\varphi \in [0, 2\pi)$ are defined as

$$\cos \theta = \frac{d_z}{|\mathbf{d}|}; \qquad\qquad e^{i\varphi} = \frac{d_x + i d_y}{\sqrt{d_x^2 + d_y^2}}. \tag{2.62}$$

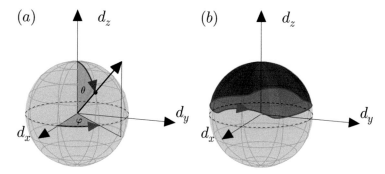

Fig. 2.2 The Bloch sphere. A generic traceless gapped two-level Hamiltonian is a linear combination of Pauli matrices, $\hat{H}(\mathbf{d}) = \mathbf{d} \cdot \hat{\sigma}$. This can be identified with a point in $\mathbb{R}^3 \backslash \{0\}$. The eigenenergies are given by the distance of the point from the origin, the eigenstates depend only on the direction of the vector $\mathbf{d}$, i.e., on the angles θ and φ, as defined in subfigure (**a**) and in Eq. (2.62) The Berry phase of a closed curve $\mathscr{C}$ is half the area enclosed by the curve when it is projected onto the surface of the Bloch sphere

We denote the two eigenstates of the Hamiltonian $\hat{H}(\mathbf{d})$ by $|+_\mathbf{d}\rangle$ and $|-_\mathbf{d}\rangle$, with

$$\hat{H}(\mathbf{d}) |\pm_\mathbf{d}\rangle = \pm|\mathbf{d}| \, |\pm_\mathbf{d}\rangle . \tag{2.63}$$

These eigenstates depend on the direction of the 3-dimensional vector $\mathbf{d}$, but not on its length. The eigenstate with $E = +|\mathbf{d}|$ of the corresponding Hamiltonian is:

$$|+_\mathbf{d}\rangle = e^{i\alpha(\theta,\varphi)} \begin{pmatrix} e^{-i\varphi/2} \cos \theta/2 \\ e^{i\varphi/2} \sin \theta/2 \end{pmatrix}, \tag{2.64}$$

while the eigenstate with $E = -|\mathbf{d}|$ is $|-_\mathbf{d}\rangle = e^{i\beta(\mathbf{d})} |+_{-\mathbf{d}}\rangle$. The choice of the phase factors α and β above corresponds to fixing a gauge. We will now review a few gauge choices.

Consider fixing $\alpha(\theta, \varphi) = 0$ for all θ, φ. This is a very symmetric choice, in this way in formula (2.64), we find $\theta/2$ and $\varphi/2$. There is problem, however, as you can see if you consider a full circle in parameter space: at any fixed value of θ, let $\varphi = 0 \to 2\pi$. We should come back to the same Hilbert space vector, and we do, but we also pick up a phase of π. We can either say that this choice of gauge led to a discontinuity at $\varphi = 0$, or that our representation is not single-valued. We now look at some attempts at fixing these problems, to find a gauge that is both continuous and single valued.

As a first attempt, let us fix $\alpha = \varphi/2$; denoting this gauge by subscript S, we have

$$|+_\mathbf{d}\rangle_S = \begin{pmatrix} \cos \theta/2 \\ e^{i\varphi} \sin \theta/2 \end{pmatrix}. \tag{2.65}$$

The phase prefactor now gives an additional factor of -1 as we make the circle in φ at fixed θ, and so it seems we have a continuous, single valued representation. There are two tricky points, however: the North Pole, $\theta = 0$, and the South Pole, $\theta = \pi$. At the North Pole, $|(0, 0, 1)\rangle_S = (1, 0)$ no problems. This gauge is problematic at the South Pole, however (which explains the choice of subscript): there, $|(0, 0, -1)\rangle_S = (0, e^{i\varphi})$, the value of the wavefunction depends on which direction we approach the South Pole from.

We can try to solve the problem at the South Pole by choosing $\alpha = -\varphi/2$, which gives us

$$|+_{\mathbf{d}}\rangle_N = \begin{pmatrix} e^{-i\varphi} \cos \theta/2 \\ \sin \theta/2 \end{pmatrix}. \tag{2.66}$$

As you can probably already see, this representation runs into trouble at the North Pole: $|(0, 0, 1)\rangle_N = (e^{-i\varphi}, 0)$.

We can try to overcome the problems at the poles by taking linear combinations of $|+_{\mathbf{d}}\rangle_S$ and $|+_{\mathbf{d}}\rangle_N$, with prefactors that vanish at the South and North Poles, respectively. A family of options is:

$$|+_{\mathbf{d}}\rangle_\chi = e^{i\chi} \sin \frac{\theta}{2} |+_{\mathbf{d}}\rangle_S + \cos \frac{\theta}{2} |+_{\mathbf{d}}\rangle_N \tag{2.67}$$

$$= \begin{pmatrix} \cos \frac{\theta}{2} (\cos \frac{\theta}{2} + \sin \frac{\theta}{2} e^{i\chi} e^{-i\varphi}) \\ \sin \frac{\theta}{2} e^{i\varphi} (\cos \frac{\theta}{2} + \sin \frac{\theta}{2} e^{i\chi} e^{-i\varphi}) \end{pmatrix}. \tag{2.68}$$

This is single valued everywhere, solves the problems at the Poles. However, it has its own problems: somewhere on the Equator, at $\theta = \pi/2$, $\varphi = \chi \pm \pi$, its norm disappears.

It is not all that surprising that we could not find a well-behaved gauge: there is none. By the end of this chapter, it should be clear, why.

2.5.2 Calculating the Berry Curvature and the Berry Phase

Consider the two-level system as defined in the previous section. Take a closed curve $\mathscr{C}$ in the parameter space $\mathbb{R}^3 \backslash \{0\}$. We are going to calculate the Berry phase γ_- of the $|-_{\mathbf{d}}\rangle$ eigenstate on this curve:

$$\gamma_-(\mathscr{C}) = \oint_{\mathscr{C}} \mathbf{A}(\mathbf{d}) d\mathbf{d}, \tag{2.69}$$

with the Berry vector potential defined as

$$\mathbf{A}(\mathbf{d}) = i \langle -_{\mathbf{d}} | \nabla_{\mathbf{d}} |-_{\mathbf{d}}\rangle. \tag{2.70}$$

The calculation becomes straightforward if we use the Berry curvature,

$$\mathbf{B}(\mathbf{d}) = \nabla_{\mathbf{d}} \times \mathbf{A}(\mathbf{d}); \tag{2.71}$$

$$\gamma_-(\mathscr{C}) = \int_{\mathscr{S}} \mathbf{B}(\mathbf{d}) d\mathscr{S}, \tag{2.72}$$

where $\mathscr{S}$ is any surface whose boundary is the loop $\mathscr{C}$. (Alternatively, it is a worthwhile exercise to calculate the Berry phase directly in a fixed gauge, e.g., one of the three gauges introduced above.)

Specifically, we make use of Berry's gauge invariant formulation (2.59) of the Berry curvature, derived in the last chapter. In the case of the generic two-level Hamiltonian (2.61), Eq. (2.59) gives

$$\mathbf{B}^{\pm}(\mathbf{d}) = -\text{Im} \frac{\langle \pm | \nabla_{\mathbf{d}} \hat{H} | \mp \rangle \times \langle \mp | \nabla_{\mathbf{d}} \hat{H} | \pm \rangle}{4\mathbf{d}^2}, \tag{2.73}$$

with

$$\nabla_{\mathbf{d}} \hat{H} = \hat{\boldsymbol{\sigma}}. \tag{2.74}$$

To evaluate (2.73), we choose the quantization axis parallel to $\mathbf{d}$, thus the eigenstates simply read

$$|+_{\mathbf{d}}\rangle = \begin{pmatrix} 1 \\ 0 \end{pmatrix}; \qquad\qquad |-_{\mathbf{d}}\rangle = \begin{pmatrix} 0 \\ 1 \end{pmatrix}. \tag{2.75}$$

The matrix elements can now be computed as

$$\langle -| \hat{\sigma}_x |+\rangle = \begin{pmatrix} 0 & 1 \end{pmatrix} \begin{pmatrix} 0 & 1 \\ 1 & 0 \end{pmatrix} \begin{pmatrix} 1 \\ 0 \end{pmatrix} = 1, \tag{2.76}$$

and similarly,

$$\langle -| \sigma_y |+\rangle = i; \tag{2.77}$$

$$\langle -| \sigma_z |+\rangle = 0. \tag{2.78}$$

So the cross product of the vectors reads

$$\langle -| \hat{\boldsymbol{\sigma}} |+\rangle \times \langle +| \hat{\boldsymbol{\sigma}} |-\rangle = \begin{pmatrix} 1 \\ i \\ 0 \end{pmatrix} \times \begin{pmatrix} 1 \\ -i \\ 0 \end{pmatrix} = \begin{pmatrix} 0 \\ 0 \\ 2i \end{pmatrix}. \tag{2.79}$$

This gives us for the Berry curvature,

$$\mathbf{B}^{\pm}(\mathbf{d}) = \pm \frac{\mathbf{d}}{|\mathbf{d}|} \frac{1}{2\mathbf{d}^2}. \tag{2.80}$$

We can recognize in this the field of a pointlike monopole source in the origin. Alluding to the analog between the Berry curvature and the magnetic field of electrodynamics (both are derived from a "vector potential") we can refer to this field, as a "magnetic monopole". Note however that this monopole exists in the abstract space of the vectors $\mathbf{d}$ and not in real space.

The Berry phase of the closed loop $\mathscr{C}$ in parameter space, according to Eq. (2.72), is the flux of the monopole field through a surface $\mathscr{S}$ whose boundary is $\mathscr{C}$. It is easy to convince yourself that this is half of the solid angle subtended by the curve,

$$\gamma_{-}(\mathscr{C}) = \frac{1}{2}\Omega_{\mathscr{C}}. \tag{2.81}$$

In other words, the Berry phase is half of the area enclosed by the image of $\mathscr{C}$, projected onto the surface of the unit sphere, as illustrated in Fig. 2.2.

What about the Berry phase of the other energy eigenstate? From Eq. (2.73), the corresponding Berry curvature $\mathbf{B}_{+}$ is obtained by inverting the order of the factors in the cross product: this flips the sign of the cross product. Therefore the Berry phases of the ground and excited state fulfil the relation

$$\gamma_{+}(\mathscr{C}) = -\gamma_{-}(\mathscr{C}). \tag{2.82}$$

One can see the same result on the Bloch sphere. Since $\langle + | - \rangle = 0$, the point corresponding to $|-\rangle$ is antipodal to the point corresponding to $|+\rangle$. Therefore, the curve traced by the $|-\rangle$ on the Bloch sphere is the inverted image of the curve traced by $|+\rangle$. These two curves have the same orientation, therefore the same area, with opposite signs.

2.5.3 Two-Band Lattice Models and Their Chern Numbers

The simplest case where a Chern number can arise is a two-band system. Consider a particle with two internal states, hopping on a two-dimensional lattice. The two internal states can be the spin of the conduction electron, but can also be some sublattice index of a spin polarized electron. In the translation invariant bulk, the wave vector $\mathbf{k} = (k_x, k_y)$ is a good quantum number, and the Hamiltonian reads

$$\hat{H}(\mathbf{k}) = \mathbf{d}(\mathbf{k})\hat{\sigma}, \tag{2.83}$$

with the function $\mathbf{d}(\mathbf{k})$ mapping each point of the Brillouin Zone to a three-dimensional vector. Since the Brillouin zone is a torus, the endpoints of the vectors $\mathbf{d}(\mathbf{k})$ map out a deformed torus in $\mathbb{R}^3 \backslash \{0\}$. This torus is a directed surface: its inside can be painted red, its outside, blue.

The Chern number of $|-\rangle$ (using the notation of Sect. 1.2, of $|u_1(\mathbf{k})\rangle$) is the flux of $\mathbf{B}_-(\mathbf{d})$ through this torus. We have seen above that $\mathbf{B}_-(\mathbf{d})$ is the magnetic field of a monopole at the origin $\mathbf{d} = 0$. If the origin is on the inside of the torus, this flux is $+1$. If it is outside of the torus, it is 0. If the torus is turned inside out, and contains the origin, the flux is -1. The torus can also intersect itself, and therefore contain the origin any number of times.

One way to count the number of times the torus contains the origin is as follows. Take any line from the origin to infinity, and count the number of times it intersects the torus, with a $+1$ for intersecting from the inside, and a -1 for intersecting from the outside. The sum is independent of the shape of the line, as long as it goes all the way from the origin to infinity.

Problems

2.1 Discrete Berry phase and Bloch vectors
Take an ordered set of three arbitrary, normalized states of a two-level system. Evaluate the corresponding discrete Berry phase. Each state is represented by a vector on the Bloch sphere. Show analytically that if two of the vectors coincide, then the discrete Berry phase vanishes.

2.2 Two-level system and the Berry connection
Consider the two-level system defined in Eq. (2.61), and describe the excited energy eigenstates using the gauge $|+_\mathbf{d}\rangle_S$ defined in Eq. (2.65). Using this gauge, evaluate and visualize the corresponding Berry connection vector field $\mathbf{A}(\mathbf{d})$. Is it well-defined in every point of the parameter space? Complete the same tasks using the gauge $|+_\mathbf{d}\rangle_N$ defined in Eq. (2.66).

2.3 Massive Dirac Hamiltonian
Consider the two-dimensional *massive Dirac Hamiltonian* $\hat{H}(k_x, k_y) = m\hat{\sigma}_z + k_x\hat{\sigma}_x + k_y\hat{\sigma}_y$, where $m \in \mathbb{R}$ is a constant and the parameter space is $\mathbb{R}^2 \ni (k_x, k_y)$. (a) Take a circular loop with radius κ in the parameter space, centered around the origin. Calculate the Berry phase associated to this loop and the ground-state manifold of the Hamiltonian: $\gamma_-(m, \kappa) = ?$. (b) Calculate the Berry connection $B_-(k_x, k_y)$ for the ground-state manifold. (c) Integrate the Berry connection for the whole parameter space. How does the result depend on m?

2.4 Absence of a continuous global gauge
In Sect. 2.5.1, we have shown example gauges for the two-level system that were not globally smooth on the parameter space. Prove that such globally smooth gauge does not exist.

2.5 Chern number of two-band models

Consider a two-band lattice model with the Hamiltonian $\hat{H}(\mathbf{k}) = \mathbf{d}(\mathbf{k}) \cdot \hat{\boldsymbol{\sigma}}$. Express the Chern number of the lower-energy band in terms of $\mathbf{d}(\mathbf{k})/|\mathbf{d}(\mathbf{k})|$.

Chapter 3
Polarization and Berry Phase

The bulk polarization of a band insulator is a tricky concept. Polarization of a neutral molecule is easily defined using the difference in centers of the negative and positive charges constituting the system. When we try to apply this simple concept to the periodic bulk of a band insulator (assuming for simplicity that the positive atom cores are immobile and localized), we meet complications. The center of the negative charges should be calculated from the electronic charge density in the fully occupied valence bands. However, all energy eigenstates in the valence band are delocalized over the bulk, and so the center of charge of each electron in such a state is ill defined. Nevertheless, insulators are polarizable, and respond to an external electric field by a rearrangement of charges, which corresponds to a (tiny) current in the bulk. Thus, there should be a way to define a bulk polarization.

In this chapter we show how a bulk polarization can be defined for band insulators using the so-called modern theory of polarization [23, 27, 28]. The contribution of the electrons to the polarization is a property of the many-body electron state, a Slater determinant of the energy eigenstates from the fully occupied valence bands. The central idea is to rewrite the same Slater determinant using a different orthonormal basis, one composed of localized states, the so-called Wannier states. The contribution of each electron in a Wannier state to the center of charge can then be easily assessed, and then added up.

We discuss the simplest interesting case, that of a one-dimensional two-band insulator with one occupied and one empty band, and leave the multiband case for later. We show that the center of charge of the Wannier states can be identified with the Berry phase of the occupied band over the Brillouin zone, also known as the Zak phase [38].

For a more complete and very pedagogical introduction to the Berry phase in electron wavefunctions, we refer the reader to a set of lecture notes by Resta [26].

© Springer International Publishing Switzerland 2016
J.K. Asbóth et al., *A Short Course on Topological Insulators*, Lecture Notes in Physics 919, DOI 10.1007/978-3-319-25607-8_3

3.1 The Rice-Mele Model

The toy model we use in this chapter is the Rice-Mele model, obtained from the SSH model of Chap. 1 by adding an extra staggered onsite potential. The Hamiltonian for the Rice-Mele model on a chain of N unit cells reads

$$\hat{H} = v \sum_{m=1}^{N} \left(|m, B\rangle \langle m, A| + h.c. \right) + w \sum_{m=1}^{N-1} \left(|m+1, A\rangle \langle m, B| + h.c. \right)$$

$$+ u \sum_{m=1}^{N} \left(|m, A\rangle \langle m, A| - |m, B\rangle \langle m, B| \right), \qquad (3.1)$$

with the staggered onsite potential u, the intracell hopping amplitude v, and intercell hopping amplitude w all assumed to be real. The matrix of the Hamiltonian for the Rice-Mele model on a chain of $N = 4$ sites reads

$$H = \begin{pmatrix} u & v & 0 & 0 & 0 & 0 & 0 & 0 \\ v & -u & w & 0 & 0 & 0 & 0 & 0 \\ 0 & w & u & v & 0 & 0 & 0 & 0 \\ 0 & 0 & v & -u & w & 0 & 0 & 0 \\ 0 & 0 & 0 & w & u & v & 0 & 0 \\ 0 & 0 & 0 & 0 & v & -u & v & 0 \\ 0 & 0 & 0 & 0 & 0 & w & u & v \\ 0 & 0 & 0 & 0 & 0 & 0 & v & -u \end{pmatrix}. \qquad (3.2)$$

3.2 Wannier States in the Rice-Mele Model

The bulk energy eigenstates of a band insulator are delocalized over the whole system. We use as an example the bulk Hamiltonian of the Rice-Mele model, i.e., the model on a ring of N unit cells. As in the case of the SSH model, Sect. 1.2, the energy eigenstates are the plane wave Bloch states,

$$|\Psi(k)\rangle = |k\rangle \otimes |u(k)\rangle, \qquad (3.3)$$

with

$$|k\rangle = \frac{1}{\sqrt{N}} \sum_{m=1}^{N} e^{imk} |m\rangle, \qquad \text{for } k \in \{\delta_k, 2\delta_k, \ldots, N\delta_k\} \quad \text{with } \delta_k = \frac{2\pi}{N}. \qquad (3.4)$$

We omit the index 1 from the eigenstate for simplicity. The $|u(k)\rangle$ are eigenstates of the bulk momentum-space Hamiltonian,

$$H(k) = \begin{pmatrix} u & v + we^{-ik} \\ v + we^{ik} & -u \end{pmatrix}, \tag{3.5}$$

with eigenvalue $E(k)$.

The Bloch states $|\Psi(k)\rangle$ are spread over the whole chain. They span the occupied subspace, defined by the projector

$$\hat{P} = \sum_{k \in BZ} |\Psi(k)\rangle \langle \Psi(k)|. \tag{3.6}$$

The phase of each Bloch eigenstate $|\Psi(k)\rangle$ can be set at will. A change of these phases, a gauge transformation,

$$|u(k)\rangle \to e^{i\alpha(k)} |u(k)\rangle ; \qquad |\Psi(k)\rangle \to e^{i\alpha(k)} |\Psi(k)\rangle , \tag{3.7}$$

gives an equally good set of Bloch states, with an arbitrary set of phases $\alpha(k) \in \mathbb{R}$ for $k = \delta_k, 2\delta_k, \ldots, 2\pi$. Using this freedom it is in principle possible to ensure that in the thermodynamic limit of $N \to \infty$, the components of $|\Psi(k)\rangle$ are smooth, continuous functions of k. However, this gauge might not be easy to obtain by numerical methods. We therefore prefer, if possible, to work with gauge-independent quantities, like the projector to the occupied subspace defined in Eq. (3.6).

3.2.1 Defining Properties of Wannier States

The Wannier states $|w(j)\rangle \in \mathscr{H}_{external} \otimes \mathscr{H}_{internal}$, with $j = 1, \ldots, N$, are defined by the following properties:

$$\langle w(j') \mid w(j) \rangle = \delta_{j'j} \qquad \text{Orthonormal set} \tag{3.8a}$$

$$\sum_{j=1}^{N} |w(j)\rangle \langle w(j)| = \hat{P} \qquad \text{Span the occupied subspace} \tag{3.8b}$$

$$\langle m + 1 \mid w(j+1) \rangle = \langle m \mid w(j) \rangle \qquad \text{Related by translation} \tag{3.8c}$$

$$\lim_{N \to \infty} \langle w(N/2)| (\hat{x} - N/2)^2 |w(N/2)\rangle < \infty \qquad \text{Localization} \tag{3.8d}$$

with the addition in Eq. (3.8c) defined modulo N. Requirement (3.8d), that of localization, uses the position operator,

$$\hat{x} = \sum_{m=1}^{N} m \left(|m, A\rangle \langle m, A| + |m, B\rangle \langle m, B| \right), \qquad (3.9)$$

and refers to a property of $|w(j)\rangle$ in the thermodynamic limit of $N \to \infty$ that is not easy to define precisely. In this one-dimensional case it can be turned into an even stricter requirement of exponential localization, $\langle w(j) | m \rangle \langle m | w(j) \rangle < e^{-|j-m|/\xi}$ for some finite localization length $\xi \in \mathbb{R}$.

3.2.2 Wannier States Are Inverse Fourier Transforms of the Bloch Eigenstates

Because of Bloch's theorem, all energy eigenstates have a plane wave form not only in the canonical basis $|m, \alpha\rangle$, with $\alpha = A, B$, but in the Wannier basis as well,

$$|\Psi(k)\rangle = e^{-i\alpha(k)} \frac{1}{\sqrt{N}} \sum_{j=1}^{N} e^{ikj} |w(j)\rangle, \qquad (3.10)$$

with some phase factors $\alpha(k)$. To convince yourself of this, consider the components of the right-hand-side in the basis of Bloch eigenstates. The right-hand-side is an eigenstate of the lattice translation operator S, with eigenvalue e^{-ik}, and therefore, orthogonal to all of the Bloch eigenstates $|\Psi(k')\rangle$ with $k' \neq k$. It is also orthogonal to all positive energy eigenstates, since it is in the occupied subspace. Thus, the only state left is $|\Psi(k)\rangle$.

From Eq. (3.10), an inverse Fourier transformation gives us a practical Ansatz for Wannier states,

$$|w(j)\rangle = \frac{1}{\sqrt{N}} \sum_{k=\delta_k}^{N\delta_k} e^{-ijk} e^{i\alpha(k)} |\Psi(k)\rangle. \qquad (3.11)$$

There is still a large amount of freedom left by this form, since the gauge function $\alpha(k)$ is unconstrained. This freedom can be used to construct Wannier states as localized as possible. If, e.g., a smooth gauge is found, where in the $N \to \infty$ limit, the components of $e^{i\alpha(k)} |\Psi(k)\rangle$ are analytic functions of k, we have exponential localization of the Wannier states due to properties of the Fourier transform. (More generally, if a discontinuity appears first in the lth derivative of $|\Psi(k)\rangle$, the components of the Wannier state $|w(j)\rangle$ will decay as $\langle m | w(j) \rangle \propto |m - j|^{-l-1}$.)

3.2.3 Wannier Centers Can Be Identified with the Berry Phase

We first assume that we have found a continuous gauge. The center of the Wannier state $|w(0)\rangle$ can be calculated, using

$$\hat{x}\,|w(0)\rangle = \frac{1}{2\pi} \int_{-\pi}^{\pi} dk \sum_{m} m e^{ikm} \, |m\rangle \otimes |u(k)\rangle$$

$$= -\frac{i}{2\pi} \left[\sum_{m} e^{ikm} \, |m\rangle \otimes |u(k)\rangle \right]_{-\pi}^{\pi} + \frac{i}{2\pi} \int_{-\pi}^{\pi} dk \sum_{m} e^{ikm} \, |m\rangle \otimes |\partial_k u(k)\rangle$$

$$= \frac{i}{2\pi} \int_{-\pi}^{\pi} dk \sum_{m} e^{ikm} \, |m\rangle \otimes |\partial_k u(k)\rangle . \qquad (3.12)$$

We find that the center of the Wannier state $|w(j)\rangle$ is

$$\langle w(j)|\,\hat{x}\,|w(j)\rangle = \frac{i}{2\pi} \int_{-\pi}^{\pi} dk \, \langle u(k)\,|\,\partial_k u(k)\rangle + j. \qquad (3.13)$$

The second term in this equation shows that the centers of the Wannier states are equally spaced, at a distance of one unit cell from each other. The first term, which is the Berry phase (divided by 2π) of the occupied band across the Brillouin zone, cf. Eq. (2.53), corresponds to a uniform displacement of each Wannier state by the same amount.

We define the bulk electric polarization to be the Berry phase of the occupied band across the Brillouin zone, the first term in Eq. (3.13),

$$P_{\text{electric}} = \frac{i}{2\pi} \int_{-\pi}^{\pi} dk \, \langle u(k)\,|\,\partial_k u(k)\rangle . \qquad (3.14)$$

Although the way we derived this above is intuitive, it remains to be shown that this is a consistent definition. From Chap. 2, it is clear that a gauge transformation can only change the bulk electric polarization by an integer. We will show explicitly in Chap. 5 that the change of this polarization in a quasi-adiabatic process correctly reproduces the bulk current.

3.2.4 Wannier States Using the Projected Position Operator

A numerically stable, gauge invariant way to find a tightly localized set of Wannier states is using the unitary position operator [28],

$$\hat{X} = e^{i\delta_k \hat{x}}. \qquad (3.15)$$

This operator is useful, because it fully respects the periodic boundary conditions of the ring. The eigensystem of $\hat{X}$ consists of eigenstates localized in cell m with eigenvalue $e^{i\delta_k m}$. Thus, we can associate the expectation value of the position in state $|\Psi\rangle$ with the phase of the expectation value of $\hat{X}$,

$$\langle x \rangle = \frac{N}{2\pi} \arg \langle \Psi | \hat{X} | \Psi \rangle . \tag{3.16}$$

The real part of the logarithm carries information about the degree of localization [1, 28].

In order to obtain the Wannier states, we restrict the unitary position operator to the filled bands, defining

$$\hat{X}_P = \hat{P}\hat{X}\hat{P}. \tag{3.17}$$

We will show below that in the thermodynamic limit of $N \rightarrow \infty$, the eigenstates of the projected position operator $\hat{X}_P$ form Wannier states.

To simplify the operator $\hat{X}_P$, consider

$$\langle \Psi(k') | \hat{X} | \Psi(k) \rangle = \frac{1}{N} \sum_{m'=1}^{N} e^{-im'k'} \langle m' | \otimes \langle u(k') | \sum_{m=1}^{N} e^{i\delta_k m} e^{imk} |m\rangle \otimes |u(k)\rangle$$

$$= \frac{1}{N} \langle u(k') | u(k) \rangle \sum_{m=0}^{N-1} e^{im(k+\delta_k-k')} = \delta_{k+\delta_k,k'} \langle u(k+\delta_k) | u(k) \rangle \tag{3.18}$$

where $\delta_{k+\delta_k,k'} = 1$ if $k' = k + \delta_k$, and 0 otherwise. Using this, we have

$$\hat{X}_P = \sum_{k'k} |\Psi(k')\rangle \langle \Psi(k') | \hat{X} | \Psi(k) \rangle \langle \Psi(k)|$$

$$= \sum_{k} \langle u(k+\delta_k) | u(k) \rangle \cdot |\Psi(k+\delta_k)\rangle \langle \Psi(k)| . \tag{3.19}$$

We can find the eigenvalues of $\hat{X}_P$ using a direct consequence of Eq. (3.19), namely, that raising $\hat{X}_P$ to the Nth power gives an operator proportional to the unity in the occupied subspace,

$$\left(\hat{X}_P\right)^N = W\hat{P}. \tag{3.20}$$

We will refer to the constant of proportionality, $W \in \mathbb{C}$, given by

$$W = \langle u(2\pi) | u(2\pi - \delta_k) \rangle \cdot \ldots \cdot \langle u(2\delta_k) | u(\delta_k) \rangle \langle u(\delta_k) | u(2\pi) \rangle , \tag{3.21}$$

as the Wilson loop. Note that W is very similar to a discrete Berry phase, apart from the fact that $|W| \leq 1$ (although $\lim_{N \to \infty} |W| = 1$). The spectrum of eigenvalues of $\hat{X}_P$ is therefore composed of the Nth roots of W,

$$\lambda_n = e^{in\delta_k + \log(W)/N}, \quad \text{with} \quad n = 1, \ldots, N; \quad \Longrightarrow \quad \lambda_n^N = W. \quad (3.22)$$

These eigenvalues have the same magnitude $|\lambda_n| = \sqrt[N]{|W|} < 1$, and phases in the interval $[0, 2\pi)$, spaced by δ_k. Because $\langle w(j)| \hat{X}_P |w(j)\rangle = \langle w(j)| \hat{X} |w(j)\rangle$, the magnitude tells us about the localization properties of the Wannier states, and the phases can be interpreted as position expectation values.

We now check whether eigenstates of $\hat{X}_P$ fulfil the properties required of Wannier states, Eq. (3.8). The relation $(\hat{X}_P)^N = W\hat{P}$ above shows that eigenvectors of $\hat{X}_P$ span the occupied subspace. Let us introduce the translation operator $\hat{S} = \sum_{m=1}^{N} |(m \bmod N) + 1\rangle \langle m| \otimes \mathbb{I}_{\text{internal}}$. The eigenstates are related by translation, since

$$\hat{S}^\dagger \hat{X}_P \hat{S} = e^{i\delta_k} \hat{X}_P; \quad (3.23)$$

$$\hat{X}_P |\Psi\rangle = |W|^{1/N} e^{i\alpha} |\Psi\rangle; \quad (3.24)$$

$$\hat{X}_P \hat{S} |\Psi\rangle = |W|^{1/N} e^{i\alpha + \delta_k} \hat{S} |\Psi\rangle. \quad (3.25)$$

There is a problem with the orthogonality of the eigenstates though. We leave the proof of localization as an exercise for the reader.

The projected unitary position operator $\hat{X}_P$ is a normal operator only in the thermodynamic limit of $N \to \infty$. For finite N, it is not normal, i.e., it does not commute with its adjoint, and as a result, its eigenstates do not form an orthonormal basis. This can be seen as a discretization error.

3.3 Inversion Symmetry and Polarization

For single-component, continuous-variable wavefunctions $\Psi(r)$, inversion about the origin (also known as parity) has the effect $\Psi(r) \to \hat{\Pi}\Psi(r) = \Psi(-r)$. Two important properties of the unitary operator $\hat{\Pi}$ representing inversion follow: $\hat{\Pi}^2 = 1$, and $\hat{\Pi} e^{ikr} = e^{-ikr}$. A Hamiltonian has inversion symmetry if $\hat{\Pi}\hat{H}\hat{\Pi}^\dagger = \hat{H}$.

When generalizing the inversion operator to lattice models of solid state physics with internal degrees of freedom, we have to keep two things in mind.

First, in a finite sample, the edges are bound to break inversion symmetry about the origin (except for very fine-tuned sample preparation). We therefore only care about inversion symmetry in the bulk, and require that it take $|k\rangle \to |-k\rangle$.

Second, each unit cell of the lattice models we consider also has its internal Hilbert space, which can be affected by inversion. This includes spin components (untouched by inversion) and orbital type variables (affected by inversion) as well.

In general, we represent the action of inversion on the internal Hilbert space by a unitary operator $\hat{\pi}$ independent of the unit cell.

The inversion operator is represented on the bulk Hamiltonian of a lattice model by an operator $\hat{\Pi}$, which acts on $\mathcal{H}_{\text{internal}}$ as $\hat{\pi}$,

$$\hat{\Pi} \, |k\rangle \otimes |u\rangle = |-k\rangle \otimes \hat{\pi} \, |u\rangle \, ; \tag{3.26}$$

$$\hat{\pi}^2 = \hat{\pi}^\dagger \hat{\pi} = \mathbb{I}_{\text{internal}}. \tag{3.27}$$

The action of the inversion operator on the bulk momentum-space Hamiltonian can be read off using its definition,

$$\hat{\Pi}\hat{H}(k)\hat{\Pi}^{-1} = \hat{\Pi} \, \langle k| \, \hat{H}_{\text{bulk}} \, |k\rangle \, \hat{\Pi}^{-1} = \langle -k| \, \hat{\Pi}\hat{H}_{\text{bulk}}\hat{\Pi}^{-1} \, |-k\rangle = \hat{\pi}\hat{H}(-k)\hat{\pi}^\dagger. \tag{3.28}$$

A lattice model has inversion symmetry in the bulk, if there exists a unitary and Hermitian $\hat{\pi}$ acting on the internal space, such that

$$\hat{\pi}\hat{H}(-k)\hat{\pi} = \hat{H}(k). \tag{3.29}$$

If all occupied bands can be adiabatically separated in energy, so we can focus on one band, with wavefunction $|u(k)\rangle$, inversion symmetry has a simple consequence. The eigenstates at $-k$ and k are related by

$$\hat{H}(k) \, |u(k)\rangle = E(k) \, |u(k)\rangle \quad \Longrightarrow \quad \hat{H}(-k)\hat{\pi} \, |u(k)\rangle = E(k)\hat{\pi} \, |u(k)\rangle \, ; \tag{3.30}$$

$$\Longrightarrow \quad |u(-k)\rangle = e^{i\phi(k)}\hat{\pi} \, |u(k)\rangle \, . \tag{3.31}$$

For the wavenumbers $k = 0$ and $k = \pi$, the so-called time-reversal invariant momenta, this says that they have states with a definite parity,

$$|u(0)\rangle = p_0 \, |u(0)\rangle \, ; \qquad\qquad |u(\pi)\rangle = p_\pi \, |u(\pi)\rangle \, , \tag{3.32}$$

$$\text{with} \quad p_0 = \pm 1; \qquad\qquad\qquad p_\pi = \pm 1. \tag{3.33}$$

3.3.1 Quantization of the Wilson Loop Due to Inversion Symmetry

We now rewrite the Wilson loop W of a band of an inversion-symmetric one-dimensional Hamiltonian, assuming we have a discretization into a number $2M$ of k-states, labeled by $j = -M + 1, \ldots, M$, as

$$|u_j\rangle = \begin{cases} |u(2\pi + j\delta_k)\rangle & \text{if } j \leq 0; \\ |u(j\delta_k)\rangle \, , & \text{otherwise.} \end{cases} \tag{3.34}$$

We use Eq. (3.31), which takes the form

$$\left|u_{-j}\right\rangle = e^{i\phi_j}\hat{\pi}\left|u_j\right\rangle. \tag{3.35}$$

The Wilson loop W of a band of an inversion symmetric one-dimensional insulator can only take on the values ± 1. We show this, using $M = 3$ as an example,

$$
\begin{aligned}
W &= \langle u_M \mid u_2\rangle \langle u_2 \mid u_1\rangle \langle u_1 \mid u_0\rangle \langle u_0 \mid u_{-1}\rangle \langle u_{-1} \mid u_{-2}\rangle \langle u_{-2} \mid u_M\rangle \\
&= \langle u_M \mid u_2\rangle \langle u_2 \mid u_1\rangle \langle u_1 \mid u_0\rangle \langle u_0 \mid e^{i\phi_1}\hat{\pi}\mid u_1\rangle \\
&\quad \langle u_1 \mid \hat{\pi} e^{-i\phi_1} e^{i\phi_2}\hat{\pi}\mid u_2\rangle \langle u_2 \mid \hat{\pi} e^{-i\phi_2}\mid u_M\rangle \\
&= \langle u_1 \mid u_0\rangle \langle u_0 \mid \hat{\pi}\mid u_1\rangle \langle u_2 \mid \hat{\pi}\mid u_M\rangle \langle u_M \mid u_2\rangle = p_0 p_\pi \tag{3.36}
\end{aligned}
$$

$$
\begin{aligned}
W &= \langle u_M \mid u_{M-1}\rangle \ldots \langle u_1 \mid u_0\rangle \langle u_0 \mid u_{-1}\rangle \ldots \langle u_{-M+1} \mid u_M\rangle \\
&= \langle u_M \mid u_{M-1}\rangle \ldots \langle u_1 \mid u_0\rangle \langle u_0 \mid e^{i\phi_1}\hat{\pi}\mid u_1\rangle \\
&\quad \langle u_1 \mid \hat{\pi} e^{-i\phi_1} e^{i\phi_2}\hat{\pi}\mid u_2\rangle \langle u_2 \mid \hat{\pi} e^{-i\phi_2} e^{i\phi_3}\hat{\pi}\mid u_3\rangle \ldots \langle u_{M+1}\mid \hat{\pi} e^{-i\phi_{M+1}}\mid u_M\rangle \\
&= \langle u_1 \mid u_0\rangle \langle u_0 \mid \hat{\pi}\mid u_1\rangle \langle u_{M-1}\mid \hat{\pi}\mid u_M\rangle \langle u_M \mid u_{M-1}\rangle = \pm 1 \tag{3.37}
\end{aligned}
$$

The statement about the Wilson loop can be translated to the bulk polarization, using Eq. (3.14). *Each band of an inversion symmetric one-dimensional insulator contributes to the bulk polarization 0 or 1/2.*

Problems

3.1 Inversion symmetry of the SSH model
Does the SSH model have inversion symmetry? If it has, then provide the corresponding local operator acting in the internal Hilbert space.

3.2 Eigenstates of the projected position operator are localized
In Sect. 3.2.4, it is claimed that the eigenstates of the projected position operator $\hat{X}_P$ form a Wannier set. One necessary condition for that statement to be true is that the eigenstates are localized, see Eq. (3.8d). Prove this.

Chapter 4
Adiabatic Charge Pumping, Rice-Mele Model

We now apply the Berry phase and the Chern number to show that by periodically and slowly changing the parameters of a one-dimensional solid, it is possible to pump particles in it. The number of particles (charge) pumped is an integer per cycle, that is given by a Chern number. Along the way we will introduce important concepts of edge state branches of the dispersion relation, and bulk–boundary correspondence. Since we are working towards understanding time-independent topological insulators, this chapter might seem like a detour. However, bulk–boundary correspondence of 2-dimensional Chern insulators, at the heart of the theory of topological insulators, is best understood via a mapping to an adiabatic charge pump.

The concrete system we use in this chapter is the simplest adiabatic charge pump, the time-dependent version of the Rice-Mele model,

$$\hat{H}(t) = v(t) \sum_{m=1}^{N} \big(|m, B\rangle \langle m, A| + h.c. \big) + w(t) \sum_{m=1}^{N-1} \big(|m+1, A\rangle \langle m, B| + h.c. \big)$$

$$+ u(t) \sum_{m=1}^{N} \big(|m, A\rangle \langle m, A| - |m, B\rangle \langle m, B| \big), \qquad (4.1)$$

with the staggered onsite potential u, intracell hopping amplitude v, and intercell hopping amplitude w all assumed to be real and periodic functions of time t. In this chapter, we are going to see how, by properly choosing the time sequences, we can ensure that particles are pumped along the chain.

© Springer International Publishing Switzerland 2016
J.K. Asbóth et al., *A Short Course on Topological Insulators*, Lecture Notes in Physics 919, DOI 10.1007/978-3-319-25607-8_4

4.1 Charge Pumping in a Control Freak Way

The most straightforward way to operate a charge pump in the Rice-Mele model is
to make sure that the system falls apart at all times to disconnected dimers. This
will happen if at any time either the intercell hopping amplitude w, or the intracell
hopping amplitude v vanishes. We can then use the staggered onsite potential to
nudge the lower energy eigenstate to the right. If during the whole cycle we keep
a finite energy difference between the two eigenstates, we can do the cycle slowly
enough to prevent excitation.

4.1.1 Adiabatic Shifting of Charge on a Dimer

As a first step towards the charge pumping protocol, consider a single dimer, i.e.,
$N = 1$. Using the adiabatic limit introduced in the last chapter, we can shift charge
from one site to the other. The Hamiltonian reads

$$\hat{H}(t) = u(t)\hat{\sigma}_z + v(t)\hat{\sigma}_x, \tag{4.2}$$

with no hopping allowed at the beginning and end of the cycle, at $t = 0$, we have
$(u, v) = (1, 0)$, we have and at $t = T$, we have $(u, v) = (-1, 0)$.

We initialize the system in the ground state, which at time $t = 0$ corresponds to
$|A\rangle$, a particle on site A. Then we switch on the hopping, which allows the particle to
spill over to site B, and once it has done that, we switch the hopping off. To ensure
that the particle spills over, we raise the onsite potential at A and lower it at B. A
practical choice is

$$u(t) = \cos(\pi t/T); \qquad\qquad v(t) = \sin(\pi t/T), \tag{4.3}$$

whereby the energy gap is at any time 2. According to the adiabatic theorem, if $H(t)$
is varied *slowly enough*, we will have shifted the charge to B at the end of the cycle.

4.1.2 Putting Together the Control Freak Sequence

Once we know how to shift a particle from $|m, A\rangle$ to $|m, B\rangle$, we can use that to shift
the particle further from $|m, B\rangle$ to $|m + 1, A\rangle$. For simplicity, we take a sequence

constructed from linear ramps of the amplitudes, using the function $f : [0, 1) \to \mathbb{R}$:

$$f(x) = \begin{cases} 8x, & \text{if } x \leq 1/8 \\ 1, & \text{if } 1/8 \leq x < 3/8 \\ 1 - 8(x - 3/8), & \text{if } 3/8 \leq x < 1/2 \\ 0, & \text{otherwise} \end{cases} \tag{4.4}$$

One period of the pump sequence, for $0 \leq t < T$, reads

$$u(t) = f(t/T) - f(t/T + \tfrac{1}{2}); \tag{4.5a}$$

$$v(t) = 2f(t/T + \tfrac{1}{4}); \tag{4.5b}$$

$$w(t) = f(t/T - \tfrac{1}{4}). \tag{4.5c}$$

This period, shown in Fig. 4.1a, is assumed to then be repeated. Note that we shifted the beginning time of the sequence: now at times $t/T = n \in \mathbb{Z}$, the Hamiltonian is the trivial SSH model, $t/T = n + 1/4$, disconnected monomers, at times $t/T = n + 1/2$, it is the nontrivial SSH model.

The time-dependent bulk momentum-space Hamiltonian reads

$$\hat{H}(k, t) = \mathbf{d}(k, t)\hat{\sigma} = (v(t) + w(t) \cos k)\hat{\sigma}_x + w(t) \sin k \hat{\sigma}_y + u(t)\hat{\sigma}_z, \tag{4.6}$$

which can be represented graphically as the path of the vector $\mathbf{d}(k, t)$ as the quasimomentum goes through the Brillouin zone, $k : 0 \to 2\pi$, for various fixed values of time t, as in Fig. 4.1b.

4.1.3 Visualizing the Motion of Energy Eigenstates

We can visualize the effects of the control freak pumps sequence in the Rice-Mele model by tracing the trajectories of the energy eigenstates. At any time t, each instantaneous energy eigenstate can be chosen confined to a single dimer: either on a single unit cell, or shared between two cells. In both cases, we can associate a position with the energy eigenstates: the expectation value of the position operator $\hat{x}$ defined as per Eq. (3.9).

The trajectories of energy eigenstates in the position-energy space, Fig. 4.2, show that the charge pump sequence works rather like a conveyor belt for the eigenstates. We engineered the sequence as a unitary operation that pushes all negative energy states in the bulk to the right at the rate of one unit cell per cycle (a current of one particle per cycle). These orthogonal states, one by one, are pushed into the right end region, which has only room for one energy eigenstate. Eigenstates cannot pile up in the right end region: if they did, this would violate unitarity of the time evolution

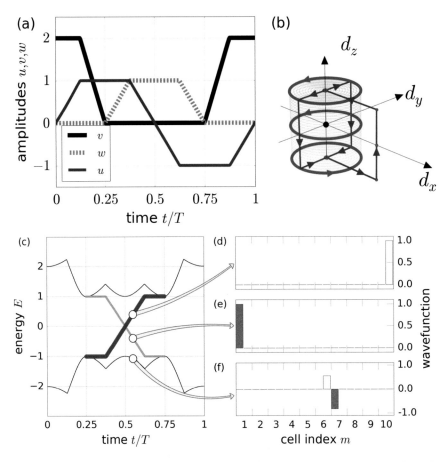

Fig. 4.1 The control freak pump sequence in the Rice-Mele model. The sequence is defined via Eqs. (4.5) and (4.6). (**a**) Time dependence of the hopping amplitudes v, w and the sublattice potential u. (**b**) The surface formed by the vector $\mathbf{d}(k,t)$ corresponding to the bulk momentum-space Hamiltonian. The topology of the surface is a torus, but its parts corresponding to $t \in [0, 0.25]T$ and $t \in [0.75, 1]T$ are infinitely thin and appear as a line due to the vanishing value of w in these time intervals. (**c**) Instantaneous spectrum of the Hamiltonian $\hat{H}(t)$ of an open chain of $N = 10$ sites. Red (blue) points represent states that are localized in the rightmost (leftmost) unit cells and have energies between -1 and 1

operator $U(t) = \mathbb{T}e^{-i\int_0^t H(t')dt'}$, where $\mathbb{T}$ stands for time ordering, since initially orthogonal states would acquire finite overlap. So, states pumped to the right edge have to go somewhere, and the only direction they can go is back towards the bulk. This on the other hand is only possible, if they acquire enough energy to be in the upper band, since all states in the lower band in the bulk are pushed towards the right. Moreover, in order to carry these states away from the right edge, and make

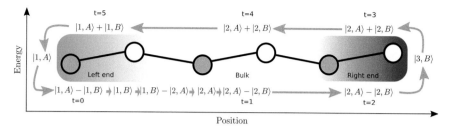

Fig. 4.2 An instantaneous energy eigenstate of the control freak pump sequence, as it is pumped through the system. At negative (positive) energy, it is pumped towards the right (left), upon reaching the right (left) end, it is pumped into the upper (lower) band

room for those coming from the bulk, the pump sequence has to push upper band bulk states towards the left.

To summarize, the control freak pump sequence is characterized by three statements. The protocol

- in the bulk, pushes all $E < 0$ eigenstates rightwards, by 1 unit cell per cycle,
- at the right end, pushes 1 eigenstate per cycle with $E < 0$ to $E > 0$,
- in the bulk, pushes all $E > 0$ eigenstates leftwards, by 1 unit cell per cycle.

If any one of these statements holds, the other two must also hold as a consequence.

4.1.4 Edge States in the Instantaneous Spectrum

We can see the charge pump at work indirectly—via its effect on the edge states—using the instantaneous spectrum, the eigenvalues of $\hat{H}(t)$ of the open chain. An example is shown for the control freak pump sequence of the Rice-Mele model on a chain of 10 unit cells (20 sites) in Fig. 4.1. Due to the special choice of the control freak sequence, the bulk consists of $N-1$-fold degenerate states (the bands are flat). More importantly, there is an energy gap separating the bands, which is open around $E = 0$ at all times. However, there are branches of the spectrum crossing this energy gap, which must represent edge states.

To assign "right" or "left" labels to edge states in the instantaneous spectrum, it is necessary to examine the corresponding wavefunctions. In case of the control freak pump sequence, right (left) edge state wavefunctions are localized on the $m = N$ ($m = 1$) unit cells, and the corresponding energy values are highlighted in green (red). The edge state branches in the dispersion in Fig. 4.1. clearly show that 1 state per cycle is pushed up in energy at the right edge.

4.2 Moving Away from the Control Freak Limit

We will now argue that the number of particles pumped by a cycle of a periodic adiabatic modulation of an insulating chain is an integer, even if the control freak attitude is relaxed. In the generic case, the energy eigenstates are delocalized over the whole bulk, and so we will need new tools to keep track of the charge pumping process. The robust quantization of charge pumping was shown by Thouless, who calculated the bulk current directly: we defer this calculation to the next chapter, and here argue using adiabatic deformations.

As an example for a generic periodically modulated insulator, we take the Rice-Mele model, but we relax the control freak attitude. We consider a smooth modulation sequence,

$$u(t) = \sin(2\pi t/T), \tag{4.7a}$$

$$v(t) = \bar{v} + \cos(2\pi t/T), \tag{4.7b}$$

$$w(t) = 1, \tag{4.7c}$$

where the sequence is fixed by choosing the average value of the intracell hopping, $\bar{v}$. With $\bar{v} = 1$, this sequence can be obtained by an adiabatic deformation of the control freak sequence. We show the smooth pump sequence and its representation in the **d** space for $\bar{v} = 1$ in Fig. 4.3.

4.2.1 Edge States in the Instantaneous Spectrum

Consider the spectrum of the instantaneous energies on an open chain, with an example for $N = 20$ unit cells shown in Fig. 4.3. Since this charge sequence was obtained by adiabatic deformation of the control freak sequence above, each branch in the dispersion relation is deformed continuously from a branch in Fig. 4.1.

The edge states are no longer confined to a single unit cell, as in the control freak case. However, as long as their energy lies deep in the bulk band gap, they have wavefunctions that decay exponentially towards the bulk, and so they can be unambiguously assigned to the left or the right end. (In case of a degeneracy between edge states at the right and left end, we might find a wavefunction with components on both ends. In that case, however, restriction of that state to the left/right end results in two separate eigenstates, to a precision that is exponentially high in the bulk length). In Fig. 4.3 we used the same simple criterion as in Chap. 1 to define edge states:

$$|\Psi\rangle \text{ is on the right edge} \iff \sum_{m=N-1}^{N} \left(|\langle\Psi \mid m, A\rangle|^2 + |\langle\Psi \mid m, B\rangle|^2 \right) > 0.6;$$

$$\tag{4.8}$$

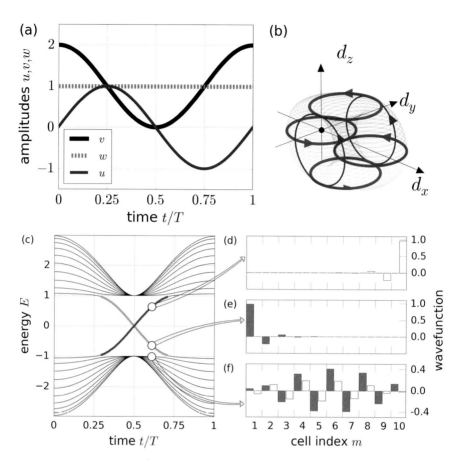

Fig. 4.3 The smooth pump sequence of the Rice-Mele model for $\bar{v} = 1$. The hopping amplitudes and the sublattice potential (**a**) are varied smoothly as a function of time. The vector $\mathbf{d}(k, t)$ corresponding to the bulk momentum-space Hamiltonian (**b**) traces out a torus in the 3-dimensional space. Instantaneous spectrum of the Hamiltonian $\hat{H}(t)$ on an open chain of $N = 10$ sites (**c**) reveals that during a cycle, one state crosses over to the upper band on the right edge, and one to the lower band on the left edge (dark red/light blue highlights energies of edge states, whose wavefunctions have than 60% weight on the rightmost/leftmost 2 unit cells). The wavefunctions of the edge states (d,e) are exponentially localized to one edge and have support overwhelmingly on one sublattice each. In contrast a typical bulk state (**f**) has a delocalized wavefunction with support on both sublattices

$$|\Psi\rangle \text{ is on the left edge} \iff \sum_{m=1}^{2} \left(|\langle \Psi \mid m, A\rangle|^2 + |\langle \Psi \mid m, B\rangle|^2 \right) > 0.6. \quad (4.9)$$

As in the control freak case, there is a branch of energy eigenstates crossing over from $E < 0$ to $E > 0$ at the right edge, and from $E > 0$ to $E < 0$, at the left.

We can define the *edge spectrum* to consist of edge state branches of the dispersion relation, that are clearly assigned to the right end. More precisely, we take two limiting energies, ε_- and ε_+, deep in the bulk gap, and only consider energy eigenstates of the open chain with eigenvalues $E_n(t)$ between these limits, $\varepsilon_- < E_n(t) < \varepsilon_+$, with eigenstates localized at the right edge. Each edge state branch can begin (1) at $t = 0$, as a continuation of another (or the same) edge state branch ending at $t = T$, or (2) at $E = \varepsilon_-$, or (3) at $E = \varepsilon_+$. Each edge state branch can end (1) at $t = T$, to then continue in another (or the same) edge state branch at $t = 0$, or (2) at $E = \varepsilon_-$, or (3) at $E = \varepsilon_+$. That is 9 possible types of edge state branches. Taking into account that the edge state spectrum, like the total spectrum, has to be periodic in t, *the number of edge state branches entering the energy range $\varepsilon_- < E < \varepsilon_+$ during a cycle is equal to the number of branches leaving it.*

4.2.2 The Net Number of Edge States Pumped in Energy Is a Topological Invariant

We now define an integer Q, that counts the number of edge states pumped up in energy across at the right edge. Although this quantity is not easily represented by a closed formula, it is straightforward to read it off from the dispersion relation of an open system. We restrict our attention to the neighbourhood of an energy ε deep in the bulk gap around $E = 0$, such that at all points where $E_n = \varepsilon$, the derivative dE_n/dt does not vanish. Then every edge state energy branch entering this neighborhood crosses $E = \varepsilon$ either towards $E > \varepsilon$ or towards $E < \varepsilon$. During one cycle, we define for the states at the right edge

$$N_+ = \text{number of times } E = \varepsilon \text{ is crossed from } E < \varepsilon \text{ to } E > \varepsilon; \qquad (4.10)$$

$$N_- = \text{number of times } E = \varepsilon \text{ is crossed from } E > \varepsilon \text{ to } E < \varepsilon; \qquad (4.11)$$

$$Q = N_+ - N_- = \text{net number of edge states pumped up in energy}. \qquad (4.12)$$

Note that within the gap, Q is independent of the choice of ε. If we found a value Q_0 at $E = \varepsilon_0$, but a different $Q_1 \neq Q_0$ at $E = \varepsilon_1 > \varepsilon_0$, this would require a net number $Q_0 - Q_1$ of edge state branches at the right edge to enter the energy region $\varepsilon_0 < E < \varepsilon_1$ during a cycle but never exit it. Since both ε_0 and ε_1 are deep in the bulk gap, away from the bulk bands, this is not possible.

The net number of edge states pumped up inside the gap on the right edge, Q, is a topological invariant: its value cannot change under continuous deformations of the Hamiltonian $H(t)$ that preserve the bulk gap. This so-called topological protection is straightforward to prove, by considering processes that might change this number. We do this using Fig. 4.4.

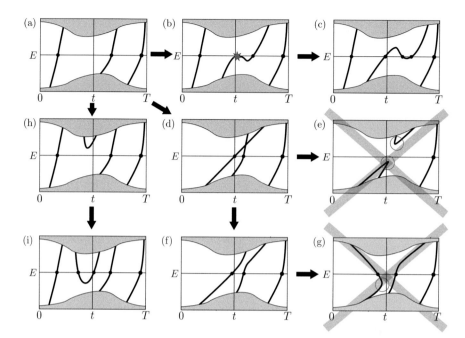

Fig. 4.4 Adiabatic deformations of dispersion relations of edge states on one edge, in an energy window that is deep inside the bulk gap. Starting from a system with 3 copropagating edge states (**a**), an edge state's dispersion relation can develop a "bump", (**b**)–(**c**). This can change the number of edge states at a given energy (intersections of the branches with the horizontal line corresponding to the energy), but always by introducing new edge states pairwise, with opposite directions of propagation. Thus the signed sum of edge states remains unchanged. Alternatively, two edge states can develop a crossing, that because of possible coupling between the edge states turns into an avoided crossing (**d**), (**f**). This cannot open a gap between branches of the dispersion relation (**e**), as this would mean that the branches become multivalued functions of the wavenumber k_x (indicating a discontinuity in $E(k_x)$, which is not possible for a system with short-range hoppings). Therefore, the signed sum of edge states is also unchanged by this process. One might think the signed sum of edge states can change if an edge state's direction of propagation changes under the adiabatic deformation, as in (**g**). However, this is also not possible, as it would also make a branch of the dispersion relation multivalued. Deformation of the Hamiltonian can also form a new edge state dispersion branch, as in (**a**)–(**h**)–(**i**), but because of periodic boundary conditions along k_x, this cannot change the signed sum of the number of edge states

The number of times edge state branches intersect $E = 0$ can change because new intersection points appear. These can form because an edge state branch is deformed, and as a result, it gradually develops a "bump", local maximum, and the local maximum gets displaced from $E < 0$ to $E > 0$. For a schematic example, see Fig. 4.4a–c. Alternatively, the dispersion relation branch of the edge state can also form a local minimum, gradually displaced from $E > 0$ to $E < 0$. In both cases, the number of intersections of the edge band with the $E = 0$ line grows by 2, but the two new intersections must have opposite pump directions. Therefore, both N_+ and N_- increase by 1, but their difference, $Q = N_+ - N_-$, stays the same.

New intersection points can also arise because a new edge state branch forms. As long as the bulk gap stays open, though, this new edge state band has to be a deformed version of one of the bulk bands, as shown in Fig. 4.4a–h–i. Because the periodic boundary conditions must hold in the Brillouin zone, the dispersion relation of the new edge state has to come from a bulk band and go back to the same bulk band, or it can be detached from the bulk band, and be entirely inside the gap. In both cases, the above argument applies, and it has to intersect the $E = 0$ line an even number of times, with no change of Q.

The number of times edge state branches intersect $E = 0$ can also decrease if two edge state branches develop an energy gap. However, to open an energy gap, the edge states have to be pumped in opposite directions. For states pumped in the same direction, energy crossing between them can become an avoided crossing, but no gap can be opened, as this would violate the single-valuedness of a dispersion relation branch, as illustrated in Fig. 4.4d–f. This same argument shows why it is not possible for an edge state to change its direction of propagation under an adiabatic deformation without developing a local maximum or minimum (which cases we already considered above). As shown in Fig. 4.4g, this would entail that at some stage during the deformation the edge state branch was not single valued.

4.3 Tracking the Charges with Wannier States

Electrons in a solid are often described via Bloch states delocalized over the whole lattice. As we have seen in Sect. 3.2 though, one can represent a certain energy band with a set of Wannier states, which inherit the spatial structure (discrete translational invariance) of the lattice, and are well localized. Therefore, it seems possible to visualize the adiabatic pumping process by following the adiabatic motion of the Wannier functions as the parameters of the lattice Hamiltonian are varied in time.

In fact, we will describe the adiabatic evolution of both the position and energy expectation values of the Wannier functions. By this, the toolbox for analyzing the adiabatic pumping procedure for control-freak-type pumping is extended to arbitrary pumping sequences.

4.3.1 Plot the Wannier Centers

According to the result (3.13), the Wannier center positions of a certain band, in units of the lattice constant, are given by the Berry phase of that band divided by 2π. Hence, to follow the motion of the Wannier centers during the pumping procedure, we calculate the Berry phase of the given band for each moment of time.

The numerically computed Wannier-center positions obtained for the smoothly modulated Rice-Mele sequence [defined via Eqs. (4.6) and (4.7)], with $\bar{v} = 1$, on a finite lattice with periodic boundary conditions, are shown in Fig. 4.5a. Solid

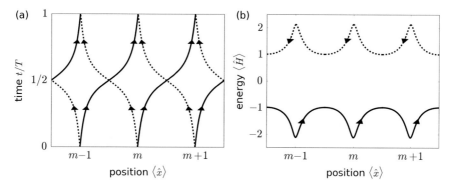

Fig. 4.5 Time evolution of Wannier centers and Wannier energies in a smoothly modulated topological Rice-Mele pumping sequence [defined via Eqs. (4.6) and (4.5)]. The parameter of the sequence is $\bar{v} = 1$, corresponding to a Chern number of 1. In both subfigures, a solid (dotted) line corresponds to the valence (conduction) band. (**a**) Time evolution of the Wannier centers of the bands. During a cycle, each Wannier center of the valence (conduction) band moves to the right (left) with a single lattice constant. (**b**) Time evolution of the position and energy expectation value of a single Wannier center in the valence/conduction band, over a few complete cycles

(dotted) lines correspond to Wannier states of the valence, i.e., lower (conduction, i.e., upper) band. The results show that during a complete cycle, each Wannier center of the valence (conduction) band moves to the right (left) with a single lattice constant. Figure 4.5b shows time evolution of the position and energy expectation value of a single Wannier center in the valence/conduction band, over a few complete cycles. In complete analogy with control-freak pumping, these results suggest that the considered pumping sequence operates as a conveyor belt: it transports valence-band electrons from left to right, with a speed of one lattice constant per cycle, and would transport conduction-band electrons, if they were present, from right to left, with the same pace.

An important result, which is not specific to the considered pumping cycle, arises from the above considerations. As pumping is cyclic, the Berry phase of a given band at $t = 0$ is equivalent to that at $t = T$. Therefore, the displacement of the Wannier center during a complete cycle is an integer.

4.3.2 Number of Pumped Particles Is the Chern Number

Our above results for the smoothly modulated Rice-Mele cycle suggest the interpretation that during a complete cycle, each electron in the filled valence band is displaced to the right by a single lattice constant. From this interpretation, it follows that the number of particles pumped from left to right, through an arbitrary cross section of the lattice, during a complete cycle, is one. Generalizing this consideration for arbitrary one-dimensional lattice models and pumping cycles, it suggests that the number of pumped particles is an integer.

We now shown that this integer is the Chern number associated to the valence band, that is, to the ground-state manifold of the time-dependent bulk momentum-space Hamiltonian $\hat{H}(k,t)$. To prove this, we first write the Wannier-center displacement $\Delta x_{0,T}$ for the complete cycle by splitting up the cycle $[0,T]$ to small pieces Δt:

$$\Delta x_{0,T} = \lim_{n\to\infty} \sum_{i=0}^{n-1} \Delta x_{t_i,t_i+\Delta t}, \tag{4.13}$$

where $\Delta t = T/n$ and $t_i = i\Delta t$. Then, we express the infinitesimal displacements with the Berry phases,

$$\Delta x(t_i, t_i + \Delta t) = \frac{i}{2\pi} \int_{-\pi}^{\pi} dk \left[\langle u_n(t_i + \Delta t) \mid \partial_k u_n(t_i + \Delta t)\rangle - \langle u_n(t_i) \mid \partial_k u_n(t_i)\rangle \right], \tag{4.14}$$

where the k argument is suppressed for brevity. Since the $k = -\pi$ and $k = \pi$ values are equivalent, the above integral can be considered as a line integral of the Berry connection to the closed boundary line ∂R_i of the infinitesimally narrow rectangle $(k,t) \in R_i = [-\pi, \pi) \times [t_i, t_i + \Delta t]$; that is,

$$\Delta x(t_i, t_i + \Delta t) = \frac{1}{2\pi} \oint_{\partial R_i} \mathbf{A}^{(n)} \cdot d\mathbf{R}. \tag{4.15}$$

Using the fact that we can choose a gauge that is locally smooth on that rectangle R_i, we obtain

$$\Delta x(t_i, t_i + \Delta t) = \frac{1}{2\pi} \int_{\partial R_i} B^{(n)} dk dt, \tag{4.16}$$

where $B^{(n)}$ is the Berry curvature associated to the nth eigenstate manifold of $\hat{H}(k,t)$. Together with Eq. (4.13), this result ensures that the Wannier-center displacement is the Chern number:

$$\Delta x_{0,T} = \frac{1}{2\pi} \int_0^T dt \int_{-\pi}^{\pi} dk B^{(n)}. \tag{4.17}$$

Note that even though we have not performed an explicit calculation of the valence-band Chern number of the smooth Rice-Mele pump cycle with $\bar{v} = 1$, by looking at the motion of the corresponding Wannier centers we can conclude that the Chern number is 1.

4.3.3 Tuning the Pump Using the Average Intracell Hopping Amplitude $\bar{v}$

So far, the discussed results were obtained for the special case of the smoothly modulated Rice-Mele pumping cycle with average intracell hopping $\bar{v} = 1$. Now we ask the question: can the number of pumped particles be changed by tuning the parameter $\bar{v}$? To show that the answer is yes, and the pump has such a tunability, on Fig. 4.6a we plot the instantaneous energy spectrum corresponding to $\bar{v} = -1$. The spectrum reveals that this sequence, similarly to the $\bar{v} = 1$ case, does pump a single particle per cycle. However, the direction of pumping is opposite in the two cases: Fig. 4.6a shows that during a cycle, one edge state on the left (light blue) crosses over from the valence band to the conduction band, revealing that the particles are pumped from right to left in the valence band.

4.3.4 Robustness Against Disorder

So far, we have the following picture of an adiabatic pump in a long open chain. If we take a cross section at the middle of the chain, a single particle will be pumped through that, from left to right, during a complete cycle. This implies that at the end of the cycle, the number of particles on the right side of the cross section has grown

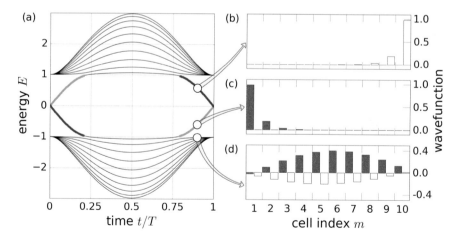

Fig. 4.6 The smooth pump sequence of the Rice-Mele model for $\bar{v} = -1$, revealing reversed pumping with respect to the $\bar{v} = 1$ case. (**a**) Instantaneous spectrum of the Hamiltonian $\hat{H}(t)$ on an open chain of $N = 10$ sites. During a cycle, one edge state on the left (light blue) crosses over from the valence to the conduction band, revealing that there is a single particle per cycle is pumped from right to left in the valence band. Wavefunctions of the edge states (**b, c**) as well as a typical bulk state (**d**) are also shown

by one, therefore the final state reached by the electron system is different from the original, ground state. This implies that during the course of the cycle, a valence-band energy eigenstate deformed into a conduction-band state, and that occurred on the right edge of the chain; the opposite happens on the left edge.

Does that qualitative behavior change if we introduce disorder in the edge regions of the open chain? No: as long as the bulk of the chain remains regular, the pump works at the middle of the chain, and therefore the above conclusion about the exchange of a pair of states between the valence and conduction bands still holds. On the other hand, introducing disorder in the bulk seems to complicate the above-developed description of pumping in terms of Wannier-center motion, and therefore might change number of edge states and the qualitative nature of the instantaneous energy spectrum.

Problems

4.1 Further control-freak pump cycles

Construct a control-freak pump cycle where the spectrum of the bulk does not change during the entire cycle, and the pumped charge is (a) zero (b) one.

Chapter 5
Current Operator and Particle Pumping

In the previous chapter, we described quantized adiabatic pumping of particles in a one-dimensional lattice in an intuitive and visual fashion, using the concepts of the control-freak pumping cycle and the time evolution of the Wannier centers. Here, we provide a more formal description of the same effect. For simplicity, we consider two-band insulator lattice models with a completely filled lower band, which are described by a periodically time-dependent bulk momentum-space Hamiltonian of the form

$$\hat{H}(k, t) = \mathbf{d}(k, t) \cdot \hat{\boldsymbol{\sigma}}, \tag{5.1}$$

where $\mathbf{d}(k, t)$ is a dimensionless three-dimensional vector fulfilling $\mathbf{d}(k, t) \geq 1$, and $\hat{\boldsymbol{\sigma}}$ is the vector of Pauli matrices. This Hamiltonian is periodic both in momentum and in time, $\hat{H}(k + 2\pi, t) = \hat{H}(k, t + T) = \hat{H}(k, t)$, where T is the period of the time dependence of the Hamiltonian. The minimal energy gap between the two eigenstates of the Hamiltonian is 2. Furthermore, the frequency characterising the periodicity of the Hamiltonian is $\Omega \equiv 2\pi/T$. We call the periodically time-dependent Hamiltonian *quasi-adiabatic*, if $\Omega \ll 1$, and the *adiabatic limit* is defined as $\Omega \to 0$, that is, $T \to \infty$.

For example, $\mathbf{d}$ can be chosen as

$$\mathbf{d}(k, t) = \begin{pmatrix} \bar{v} + \cos \Omega t + \cos k \\ \sin k \\ \sin \Omega t \end{pmatrix}, \tag{5.2}$$

corresponding to the smoothly modulated Rice-Mele model, see Eq. (4.7) and Eq. (4.6).

We will denote the eigenstate of $\hat{H}(k, t)$ with a lower (higher) energy eigenvalue as $|u_1(k, t)\rangle$ ($|u_2(k, t)\rangle$). With this notation, we can express the central result of this chapter: in adiabatic pumping, the momentum- and time-resolved current carried

© Springer International Publishing Switzerland 2016
J.K. Asbóth et al., *A Short Course on Topological Insulators*, Lecture Notes in Physics 919, DOI 10.1007/978-3-319-25607-8_5

by the electrons of the filled band is a sum of two terms: a group-velocity term
and the Berry curvature of that band. The group-velocity term does not contribute to
pumping. As a consequence, the number $\mathcal{Q}$ of particles pumped through an arbitrary
cross section of an infinite one-dimensional crystal during a complete adiabatic
cycle is the momentum- and time integral of the Berry curvature, that is,

$$\mathcal{Q} = -i\frac{1}{2\pi} \int_0^T dt \int_{-\pi}^{\pi} dk \left(\partial_k \langle u_1(k,t) | \partial_t u_1(k,t) \rangle - \partial_t \langle u_1(k,t) | \partial_k u_1(k,t) \rangle \right). \quad (5.3)$$

This is the Chern number associated to the ground-state manifold of $\hat{H}(k,t)$. As the
latter is an integer, the number of pumped particles is quantized. This result was
discovered by David Thouless [33].

We derive Eq. (5.3) via the following steps. In Sect. 5.1, we consider a generic
time-dependent lattice Hamiltonian, and we express the number of particles moving
through a cross section of the lattice using the current operator and the time-evolving
states of the particles. Then, in Sect. 5.2.2, we provide a description of the time-
evolving states in the case of periodic and quasi-adiabatic time dependence of the
lattice Hamiltonian. This allows us to express the number of pumped particles
for quasi-adiabatic time dependence. Finally, in Sect. 5.3, building on the latter
result for quasi-adiabatic pumping, we take the adiabatic limit and thereby establish
the connection between the current, the Berry curvature, the number of pumped
particles, and the Chern number.

5.1 Particle Current at a Cross Section of the Lattice

Our aim here is to express the number of particles pumped through a cross section
of the lattice, assuming that the time evolution of the Bloch states due to the
time-dependence of the Hamiltonian is known. As intermediate steps toward this
end, we derive the real-space current operator and the diagonal matrix elements
of the momentum-space current operator, and establish an important relation
between those diagonal matrix elements and the momentum-space Hamiltonian.
For concreteness, we first discuss these using the example of the Rice-Mele model
introduced in the preceding chapter. It is straightforward to generalize the results
for lattice models with a generic internal degree of freedom; the generalized results
are also given below. Finally, we use the relation between the current and the
Hamiltonian to express the number of pumped particles with the time-evolving
states and the Hamiltonian.

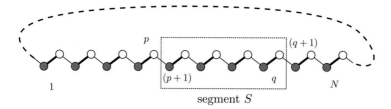

Fig. 5.1 A segment enclosed by two cross sections in the one-dimensional SSH model. The segment S is defined as the part of the chain between the pth and $(q+1)$th unit cells. The current operators corresponding to the two cross sections can be established by considering the temporal change of the number of particles in the segment. The dashed line represents the periodic boundary condition

5.1.1 Current Operator in the Rice-Mele Model

We start things off with an example and consider the Rice-Mele model with $N \gg 1$ unit cells and periodic boundary conditions (Fig. 5.1). The real-space bulk Hamiltonian $\hat{H}_{\text{bulk}}$ has almost the same form as Eq. (3.1), with the difference that the sum corresponding to intercell hopping runs up to N, and in accordance with the periodic boundary condition, the unit cell index m should be understood as $(m \bmod N)$. The bulk momentum-space Hamiltonian $\hat{H}(k)$ of the model is given in Eq. (4.6).

5.1.1.1 Influx of Particles into a Segment of the Crystal

We aim at establishing the operator representing the particle current flowing through a cross section of the one-dimensional crystal. We take a cross section between the pth and $(p+1)$th unit cells, and denote the corresponding current operator as $\hat{j}_{p+1/2}$. To find the current operator, we first consider a segment S of the crystal, stretching between (and including) the $(p+1)$th and qth unit cells, where $q \geq p+1$. The number of particles in that segment S is represented by the operator

$$\hat{\mathcal{Q}}_S \equiv \sum_{m \in S} \sum_{\alpha \in \{A,B\}} |m, \alpha\rangle \langle m, \alpha| . \tag{5.4}$$

Now, the time evolution of the number of particles embedded in the segment S follows

$$\partial_t \langle \hat{\mathcal{Q}}_S \rangle_t = -i \langle [\hat{\mathcal{Q}}_S, \hat{H}(t)] \rangle_t \tag{5.5}$$

Hence, we identify the operator describing the influx of particles into the segment as

$$\hat{j}_S(t) = -i[\hat{\mathcal{Q}}_S, \hat{H}(t)]. \tag{5.6}$$

A straightforward calculation shows that Eq. (5.6) implies

$$
\hat{j}_S(t) = -iw(t) \left(|p+1, A\rangle \langle p, B| - |p, B\rangle \langle p+1, A| \right.
$$
$$
\left. + |q, B\rangle \langle q+1, A| - |q+1, A\rangle \langle q, B| \right) \tag{5.7}
$$

Remarkably, the operator $\hat{j}_S(t)$ is time dependent, if the hopping amplitude $w(t)$ is time dependent.

5.1.1.2 Current Operator at a Cross Section of the Crystal

Clearly, the terms in Eq. (5.7) can be separated into two groups: the first two terms are hopping operators bridging the cross section $p + 1/2$ (that is, the cross section between unit cell p and unit cell $p + 1$), and the last two terms are bridging the cross section $q + 1/2$. Thereby we define

$$
\hat{j}_{m+1/2}(t) \equiv -iw(t) \left(|m+1, A\rangle \langle m, B| - |m, B\rangle \langle m+1, A| \right), \tag{5.8}
$$

and use this definition to express $\hat{j}_S$ as

$$
\hat{j}_S = \hat{j}_{p+1/2} - \hat{j}_{q+1/2}. \tag{5.9}
$$

This relation allows us to interpret $\hat{j}_{m+1/2}(t)$ as the current operator describing particle flow, from left to right, across the cross section $m + 1/2$.

5.1.1.3 Relation of the Current Operator to the Hamiltonian and to the Group Velocity

Later we will need the momentum-diagonal matrix elements of the current operator, which are defined as

$$
\hat{j}_{m+1/2}(k, t) = \langle k | \hat{j}_{m+1/2}(t) | k \rangle. \tag{5.10}
$$

For the Rice-Mele model under consideration, these can be expressed using Eqs. (1.8) and (5.8) as

$$
\hat{j}_{m+1/2}(k, t) = \frac{1}{N} \begin{pmatrix} 0 & -iw(t)e^{-ik} \\ iw(t)e^{ik} & 0 \end{pmatrix}. \tag{5.11}
$$

From a comparison of this result and Eq. (4.6), we see that the momentum-diagonal matrix elements of the current operator are related to the momentum-space Hamiltonian as

$$
\hat{j}_{m+1/2}(k, t) = \frac{1}{N} \partial_k \hat{H}(k, t). \tag{5.12}
$$

This is the central result of this section. Even though we have derived it only for the case of the Rice-Mele model, it is a generic result. A generalization is outlined in the next section.

In the case of a lattice without an internal degree of freedom, it is easy to see that Eq. (5.12) establishes the equivalence $j_{m+1/2}(k,t) = v_k(t)/N$ between the current and the (instantaneous) group velocity $v_k(t)$: in this case, the bulk momentum-space Hamiltonian $\hat{H}(k,t)$ equals the dispersion relation $E(k,t)$, and its momentum derivative $v_k(t) = \frac{\partial E(k,t)}{\partial k}$ is defined as the group velocity of the energy eigenstates. This correspondence generalizes to lattices with an internal degree of freedom as well. The current carried by an instantaneous energy eigenstate $|\Psi_n(k,t)\rangle = |k\rangle \otimes |u_n(k,t)\rangle$ of such a lattice is

$$
\begin{aligned}
\langle u_n| \hat{j}_{m+1/2} |u_n\rangle &= \frac{1}{N} \langle u_n| \left[\partial_k \hat{H} \right] |u_n\rangle \\
&= \frac{1}{N} \langle u_n| \left[\partial_k \sum_{n'} E_{n'} |u_{n'}\rangle \langle u_{n'}| \right] |u_n\rangle \\
&= \frac{1}{N} \{ \langle u_n| \left[(\partial_k E_n) |u_n\rangle \langle u_n| + E_n |\partial_k u_n\rangle \langle u_n| + E_n |u_n\rangle \langle \partial_k u_n| \right] |u_n\rangle \} \\
&= \frac{1}{N} (\partial_k E_n + E_n \partial_k \langle u_n| u_n\rangle) = \frac{\partial_k E_n}{N} = \frac{v_{n,k}}{N},
\end{aligned}
\tag{5.13}
$$

where the arguments (k,t) are suppressed for brevity.

5.1.2 Current Operator in a Generic One-Dimensional Lattice Model

Here we prove Eq. (5.12) in a more general setting. Previously, we focused on the Rice-Mele model that has only two bands and hopping only between nearest-neighbor cells. Consider now a general one-dimensional lattice model with N_b bands and finite-range hopping with range $1 \le r \ll N$; here, $r = 1$ corresponds to hopping between neighbouring unit cells only. As before, we take a long chain with $N \gg 1$ unit cells, and assume periodic boundary conditions.

The real-space Hamiltonian has the form

$$
\hat{H}(t) = \sum_{m,m'=1}^{N} \sum_{\alpha,\alpha'=1}^{N_b} H_{m\alpha,m'\alpha'}(t) |m,\alpha\rangle \langle m'\alpha'|,
\tag{5.14}
$$

where m and m' are unit cell indices, and $\alpha, \alpha' \in \{1, 2, \ldots, N_b\}$ correspond to the internal degree of freedom within the unit cell. Due to the finite-range-hopping assumption, the Hamiltonian can also be written as

$$\hat{H}(t) = \sum_{m=1}^{N} \sum_{i=-r}^{r} \sum_{\alpha,\alpha'=1}^{N_b} H_{m+i,\alpha;m,\alpha'}(t) \, |m+i,\alpha\rangle \langle m\alpha'| . \tag{5.15}$$

Again, a unit cell index m should be understood as $(m \mod N)$. Also, due to the discrete translational invariance, we have $H_{m+i,\alpha;m,\alpha'} = H_{i\alpha;0,\alpha'}$, implying

$$\hat{H}(t) = \sum_{m=1}^{N} \sum_{i=-r}^{r} \sum_{\alpha,\alpha'=1}^{N_b} H_{i,\alpha;0\alpha'}(t) \, |m+i,\alpha\rangle \langle m\alpha'| \tag{5.16}$$

The bulk momentum-space Hamiltonian then reads

$$\hat{H}(k,t) \equiv \langle k| \hat{H}(t) |k\rangle = \sum_{m=-r}^{r} H_{m,\alpha;0,\alpha'}(t) e^{-ikm} \, |\alpha\rangle \langle \alpha'| . \tag{5.17}$$

Next, we establish the operator representing the particle current flowing through a cross section of the one-dimensional crystal, the same way we did in Sect. 5.1.1. We take a cross section between the pth and $(p+1)$th unit cells, and denote the corresponding current operator as $\hat{j}_{p+1/2}$. To find the current operator, we first consider a segment S of the crystal, stretching between (and including) the $(p+1)$th and qth unit cells, where $q - p \geq r$. The number of particles embedded in that long segment is represented by the operator

$$\hat{\mathcal{Q}}_S \equiv \sum_{m \in S} \sum_{\alpha=1}^{N_b} |m\alpha\rangle \langle m\alpha| . \tag{5.18}$$

As discussed in the preceding section, we identify the operator describing the influx of particles into the wire segment as

$$\hat{j}_S(t) = -i[\hat{\mathcal{Q}}_S, \hat{H}(t)]. \tag{5.19}$$

From this, a straightforward calculation shows that

$$\hat{j}_S(t) = -i \sum_{m \in S} \sum_{m' \notin S} \sum_{\alpha,\alpha'=1}^{N_b} \left[H_{m\alpha,m'\alpha'}(t) \, |m\alpha\rangle \langle m'\alpha'| \right.$$
$$\left. - H_{m'\alpha',m\alpha}(t) \, |m'\alpha'\rangle \langle m\alpha| \right]. \tag{5.20}$$

Note that Eq. (5.20) testifies that the operator $\hat{j}_S(t)$ is constructed only from those hopping matrix elements of the Hamiltonian that bridge either the $p+1/2$ or the $q+1/2$ cross sections of the crystal, i.e., one of the two cross sections that terminate the segment under consideration. This is ensured by the condition that the segment is at least as long as the range r of hopping. A further consequence of this is that the terms in Eq. (5.20) can be separated into two groups: one containing the hopping matrix elements bridging the cross section $p + 1/2$, and one with those bridging the cross section $q + 1/2$. The former reads

$$\hat{j}_{p+1/2}(t) = -i \sum_{m=p+1}^{p+r} \sum_{m'=p+1-r}^{p} \sum_{\alpha,\alpha'=1}^{N_b} \left[H_{m\alpha,m'\alpha'}(t) \, |m\alpha\rangle \langle m'\alpha'| \right.$$
$$\left. - H_{m'\alpha',m\alpha}(t) \, |m'\alpha'\rangle \langle m\alpha| \right].$$

(5.21)

Using this as a definition for any cross section $m + 1/2$, we conclude that Eq. (5.9) holds without any change in this generalized case as well. This conclusion allows us to interpret $\hat{j}_{m+1/2}$ as the current operator describing particle flow, from left to right, across the cross section $m + 1/2$.

After defining the momentum-diagonal matrix elements of the current operator exactly the same way as in Eq. (5.10), the relation between the current operator and the Hamiltonian has exactly the same form as in Eq. (5.12). This can be proven straightforwardly using Eqs. (5.10), (5.21), and the k-derivative of Eq. (5.17).

5.1.3 Number of Pumped Particles

Let us return to particle pumping in insulating two-band models. As the electrons are assumed to be non-interacting, and the time-dependent lattice Hamiltonian has a discrete translational invariance for all times, the many-electron state $\Phi(t)$ is a Slater determinant of Bloch-type single-particle states $|\tilde{\Psi}_1(k, t)\rangle = |k\rangle \otimes |\tilde{u}_1(k, t)\rangle$. Here, we adopted the notation introduced in Sect. 1.2, with $n = 1$ referring to the filled band. There are also two additions with respect to the notation of Sect. 1.2: first, we explicitly denote the time dependence of the state; second, we added a tilde here to denote that the state $|\tilde{\Psi}_1(k, t)\rangle$ is not an instantaneous lower-band eigenstate $|\Psi_1(k, t)\rangle$ of the Hamiltonian, but is slightly different from that due to the quasi-adiabatic driving.

The number of particles pumped through the cross section $m + 1/2$ within the time interval $t \in [0, T]$ is the time-integrated current, that is,

$$\mathcal{Q} = \int_0^T dt \, \langle \Phi(t) | j_{m+1/2}^M(t) | \Phi(t) \rangle,$$

(5.22)

where $j_{m+1/2}^M(t)$ is the many-particle generalisation of the current operator defined in Eq. (5.21), or, for the special case of the Rice-Mele model, in Eq. (5.8), and $\Phi(t)$ is the many-electron Slater determinant formed by the filled Bloch-type single-particle states, introduced in the preceding paragraph. Equation (5.22) can be converted to an expression with single-particle states:

$$\mathscr{Q} = \int_0^T dt \sum_{k \in \text{BZ}} \langle \tilde{\Psi}_1(k, t) | \hat{j}_{m+1/2}(t) | \tilde{\Psi}_1(k, t) \rangle, \qquad (5.23)$$

which is related to the momentum-diagonal matrix elements of the current operator as

$$\mathscr{Q} = \int_0^T dt \sum_{k \in \text{BZ}} \langle \tilde{u}_1(k, t) | \hat{j}_{m+1/2}(k, t) | \tilde{u}_1(k, t) \rangle. \qquad (5.24)$$

Here, BZ stands for Brillouin Zone. Finally, this is rewritten using the current-Hamiltonian relation Eq. (5.12) as

$$\mathscr{Q} = \frac{1}{N} \int_0^T dt \sum_{k \in \text{BZ}} \langle \tilde{u}_1(k, t) | \, \partial_k \hat{H}(k, t) | \tilde{u}_1(k, t) \rangle. \qquad (5.25)$$

To evaluate this in the case of adiabatic, periodically time-dependent Hamiltonian, we first need to understand how the two-level wave functions $|\tilde{u}_1(k, t)\rangle$ evolve in time in the quasi-adiabatic case; then we can insert those in Eq. (5.25), and take the adiabatic limit.

5.2 Time Evolution Governed by a Quasi-Adiabatic Hamiltonian

Our goal here is to describe the time evolution of Bloch-type electronic energy eigenstates. Nevertheless, as the pumping dynamics preserves the wavenumber k, the task simplifies to describe the dynamics of distinct two-level systems, labelled by the wavenumber k. Therefore, in this section we discuss the dynamics of a single two-level system, hence the wavenumber k does not appear in the formulas. We will restore k when evaluating the number of pumped particles in the next section. To describe the time evolution of the electronic states subject to quasi-adiabatic driving, it is convenient to use the so-called *parallel-transport gauge* or *parallel-transport time parametrization*, which we introduce below. Then, our goal is reached by performing perturbation theory in the small frequency $\Omega \ll 1$ characterizing the quasi-adiabatic driving.

5.2.1 The Parallel-Transport Time Parametrization

As mentioned earlier, the instantaneous energy eigenstates of the bulk momentum-space Hamiltonian $\hat{H}(k, t)$ are denoted as $|u_n(k, t)\rangle$. Here, in order to simplify the derivations, we will use a special time parametrization (gauge) for these eigenstates, which is called the parallel-transport time parametrization or parallel-transport gauge. As mentioned above, we suppress the momentum k.

We will call the smooth time parametrization $|u_n(t)\rangle$ of the instantaneous nth eigenstate of the Hamiltonian $\hat{H}(t)$ a parallel-transport time parametrization, if for any time point t and any band n, it holds that

$$\langle u_n(t)| \, \partial_t \, |u_n(t)\rangle = 0. \tag{5.26}$$

Using a time parametrization with this property will simplify the upcoming calculations of this section.

In Sect. 2.3, we used smooth parametrizations that were defined via a parameter space, and therefore were cyclic. For any time parametrization $|u'_n(t)\rangle$ having those properties, we can construct a parallel-transport time parametrizaton $|u_n(t)\rangle$ via the definition

$$|u_n(t)\rangle = e^{i\gamma_n(t)} \, |u'_n(t)\rangle, \tag{5.27}$$

where $\gamma_n(t)$ is the adiabatic phase associated to the adiabatic time evolution of the initial state $|u'_n(t = 0)\rangle$, governed by our adiabatically varying Hamiltonian $\hat{H}(t)$:

$$\gamma_n(t) = i \int_0^t dt' \, \langle u'_n(t')| \, \partial_{t'} \, |u'_n(t')\rangle. \tag{5.28}$$

The fact that $|u_n(t)\rangle$ indeed fulfils Eq. (5.26) can be checked by performing the time derivation and the scalar product on the left hand side of the latter.

As an interpretation of Eq. (5.27), we can say that a parallel-transport time parametrization is an adiabatically time-evolving state divided by the dynamical phase factor. Furthermore, as the Berry phase factor $e^{i\gamma_n(T)}$ is, in general, different from 1, the parallel-transport time parametrization (5.27) is, in general, not cyclic.

5.2.2 Quasi-Adiabatic Evolution

Here, following Thouless [33], we describe the quasi-adiabatic time evolution using *stationary states*, also known as *Floquet states*, that are characteristic of periodically driven quantum systems. Again, we focus on two-level systems as introduced in Eq. (5.1), and suppress the wave number k in our notation. The central result of this section is Eq. (5.42), which expresses how the instantaneous ground state mixes weakly with the instantaneous excited state due to the quasi-adiabatic time dependence of the Hamiltonian. In the next section, this result is used to evaluate the particle current and the number of pumped particles.

As an example, we can consider the state corresponding to the wavenumber $k = 0$ in the smoothly modulated Rice-Mele model with $\bar{v} = 1$, see Eq. (5.2):

$$\mathbf{d}(t) = \begin{pmatrix} 2 + \cos(\Omega t) \\ 0 \\ \sin(\Omega t) \end{pmatrix}. \tag{5.29}$$

5.2.2.1   Stationary States of Periodically Driven Dynamics

The stationary states are special solutions of the periodically time-dependent Schrödinger equation, which are essentially periodic with period T; that is, which fulfill $|\psi(t + T)\rangle = e^{-i\phi} |\psi(t)\rangle$ for any t, with ϕ being a t-independent real number. The number of such nonequivalent solutions equals the dimension of the Hilbert space of the quantum system, i.e., there are two of them for the case we consider. Here we describe stationary states in the quasi-adiabatic case, when the time evolution of the Hamiltonian is slow compared to the energy gap between the instantaneous energy eigenvalues: $\Omega \ll 1$. This condition suggest that the deviation from the adiabatic dynamics is small, and therefore each stationary state is in the close vicinity of either the instantaneous ground state or the instantaneous excited state. Thereby, we will label the stationary states with the band index n, and denote them as $|\tilde{u}_n(t)\rangle$.

Since, after all, we wish to describe pumping in a lattice with a filled lower band and an empty upper band, we mostly care about the stationary state corresponding to the lower band, $|\tilde{u}_1(t)\rangle$, and therefore want to solve the time-dependent Schrödinger equation

$$- i\partial_t |\tilde{u}_1(t)\rangle + \hat{H}(t) |\tilde{u}_1(t)\rangle = 0. \tag{5.30}$$

5.2.2.2   Making Use of the Parallel-Transport Gauge

We characterize the time evolution of the wave function $|\tilde{u}_1(t)\rangle$ by a time-dependent linear combination of the instantaneous energy eigenstates:

$$|\tilde{u}_1(t)\rangle = a_1(t)e^{-i\int_0^t dt' E_1(t')} |u_1(t)\rangle + a_2(t)e^{-i\int_0^t dt' E_2(t')} |u_2(t)\rangle, \tag{5.31}$$

Recall that we are using the parallel-transport time parametrization, having the properties (5.26) and (5.27). Therefore, in the adiabatic limit $\Omega \to 0$, we already now that $a_1(t) = 1$ and $a_2(t) = 0$. Here, we are mostly interested in the quasi-adiabatic case defined via $\Omega \ll 1$, and then it is expected that $a_1(t) \sim 1$ and $a_2(t) \sim \Omega \ll 1$.

Before making use of that consideration in the form of perturbation theory in Ω, we convert the time-dependent Schrödinger equation (5.30) to two differential equations for the two unknown functions $a_1(t)$ and $a_2(t)$. We insert $|\tilde{u}_1(t)\rangle$ of Eq. (5.31) to the time-dependent Schrodinger equation (5.30), yielding

$$-i\partial_t \left[a_1(t)e^{-i\int_0^t dt' E_1(t')} |u_1(t)\rangle + a_2(t)e^{-i\int_0^t dt' E_2(t')} |u_2(t)\rangle \right] \qquad (5.32)$$

$$+\hat{H}(t) \left[a_1(t)e^{-i\int_0^t dt' E_1(t')} |u_1(t)\rangle + a_2(t)e^{-i\int_0^t dt' E_2(t')} |u_2(t)\rangle \right] = 0$$

After evaluating the time derivatives, the left hand side consists of 8 terms. Using the instantaneous eigenvalue relations $\hat{H}(t)|u_n(t)\rangle = E_n(t)|u_n(t)\rangle$, two pairs of terms annihilate each other, and only 4 terms remain:

$$\dot{a}_1(t)|u_1(t)\rangle + a_1(t)\partial_t |u_1(t)\rangle + \dot{a}_2(t)e^{-i\int_0^t dt' E(t')} |u_2(t)\rangle$$

$$+ a_2(t)e^{-i\int_0^t dt' E(t')} \partial_t |u_2(t)\rangle = 0, \qquad (5.33)$$

where $E(t) = E_2(t) - E_1(t) = 2d(t)$.

Projecting Eq. (5.33) onto $\langle u_1(t)|$ and $\langle u_2(t)|$, respectively, and making use of the parallel-transport gauge, yields

$$\dot{a}_1(t) + a_2(t)e^{-i\int_0^t dt' E(t')} \langle u_1(t)| \partial_t |u_2(t)\rangle = 0, \qquad (5.34)$$

$$a_1(t) \langle u_2(t)| \partial_t |u_1(t)\rangle + \dot{a}_2(t)e^{-i\int_0^t dt' E(t')} = 0. \qquad (5.35)$$

The latter result can be rewritten as

$$\dot{a}_2(t) = -a_1(t) \langle u_2(t)| \partial_t |u_1(t)\rangle e^{i\int_0^t dt' E(t')} \qquad (5.36)$$

5.2.2.3 Making Use of the Quasi-Adiabatic Condition

Note that the quasi-adiabatic condition has not been invoked so far; this is the next step. As mentioned above, the quasi-adiabatic nature of the Hamiltonian suggests that one of the two stationary states will be close to the instantaneous ground state, suggesting $a_1(t) \sim 1$ and $a_2(t) \sim \Omega$. Furthermore, we know that $\langle u_1(t)| \partial_t |u_2(t)\rangle \sim \Omega$, since variations in $|u_n(t)\rangle$ become slower as the adiabatic limit is approached. The latter relation is explicitly demonstrated by the example in Eq. (5.29): if we use $|u_1(t)\rangle = (-\sin(\theta/2), \cos(\theta/2))^T$ and $|u_2(t)\rangle = (\cos(\theta/2), \sin(\theta/2))^T$ with $\theta = \arctan\left(\frac{2+\cos\Omega t}{\sin\Omega t}\right)$, fulfilling the parallel-gauge criterion, we find $\langle u_1(t)| \partial_t |u_2(t)\rangle = -\Omega \frac{1+2\cos\Omega t}{10+8\cos\Omega t}$.

As we are interested in the quasi-adiabatic case $\Omega \ll 1$, we drop those terms from (5.34) and (5.36) that are at least second order in Ω. This results in

$$\dot{a}_1(t) = 0, \qquad (5.37)$$

$$\dot{a}_2(t) = -a_1(t) \langle u_2(t)| \partial_t |u_1(t)\rangle e^{i\int_0^t dt' E(t')}. \qquad (5.38)$$

If we assume $a_1(t = 0) = 1 + o(\Omega)$, then the first equation guarantees that $a_1(t) = 1 + o(\Omega)$. Then this allows for a further simplification of Eq. (5.38):

$$\dot{a}_2(t) = - \langle u_2(t)| \, \partial_t \, |u_1(t)\rangle \, e^{i \int_0^t dt' \, E(t')}. \tag{5.39}$$

5.2.2.4   Solution of the Equation of Motion

The remaining task is to solve Eq. (5.39) for $a_2(t)$. Instead of doing this in a constructive fashion, we give the solution $a_2(t)$ and prove that it indeed fulfills Eq. (5.39) up to the desired order. The solution reads

$$a_2(t) = i \frac{\langle u_2(t)| \, \partial_t \, |u_1(t)\rangle}{E(t)} e^{i \int_0^t dt' \, E(t')}. \tag{5.40}$$

First, let us check if it solves the differential equation (5.39):

$$\partial_t a_2(t) = i \frac{(\partial_t \langle u_2(t)| \, \partial_t \, |u_1(t)\rangle)}{E(t)} e^{i \int_0^t dt' \, E(t')} - i \frac{(\partial_t E(t)) \langle u_2(t)| \, \partial_t \, |u_1(t)\rangle}{E(t)^2} e^{i \int_0^t dt' \, E(t')}$$

$$- \langle u_2(t)| \, \partial_t \, |u_1(t)\rangle \, e^{i \int_0^t dt' \, E(t')}. \tag{5.41}$$

The first two terms on the right hand side scale as Ω^2, whereas the third one scales as Ω. Hence we conclude that in the quasi-adiabatic case, Eq. (5.40) is the solution of Eq. (5.39) we were after. The corresponding solution of the time-dependent Schrödinger equation (5.30) is constructed using Eqs. (5.31), $a_1(t) = 1$ and (5.38), and reads

$$|\tilde{u}_1(t)\rangle = e^{-i \int_0^t dt' \, E_1(t')} \left[|u_1(t)\rangle + i \frac{\langle u_2(t)| \, \partial_t \, |u_1(t)\rangle}{E(t)} |u_2(t)\rangle \right]. \tag{5.42}$$

In words, Eq. (5.42) assures that the stationary state has most of its weight in the instantaneous ground state $|u_1(t)\rangle$, with a small, $\sim \Omega \ll 1$ admixture of the instantaneous excited state $|u_2(t)\rangle$. Interestingly, even though this small admixture vanishes in the adiabatic limit $\Omega \to 0$, the corresponding contribution to the number of pumped particles can give a finite contribution, as the cycle period T goes to infinity in the adiabatic limit. This will be shown explicitly in the next section.

Finally, we show that this state $|\tilde{u}_1(t)\rangle$ is indeed stationary. That is proven if we can prove that $|\tilde{u}_1(T)\rangle$ is equal to $|\tilde{u}_1(0)\rangle$ up to a phase factor. This arises as the consequence of the following fact. If the Berry phase associated to the state $|u_1\rangle$ is γ, that is, if $|u_1(T)\rangle = e^{i\gamma} |u_1(0)\rangle$, then

$$[\partial_t |u_1(t)\rangle]_T = \lim_{\epsilon \to 0} \frac{|u_1(T + \epsilon)\rangle - |u_1(T)\rangle}{\epsilon} = \lim_{\epsilon \to 0} \frac{e^{i\gamma} |u_1(\epsilon)\rangle - e^{i\gamma} |u_1(0)\rangle}{\epsilon}$$

$$= e^{i\gamma} \partial_t |u_1(0)\rangle. \tag{5.43}$$

Therefore, the two terms in the square bracket of Eq. (5.42) acquire the same phase factor $e^{i\gamma}$ at the end of the cycle, hence the obtained $|\tilde{u}_1(t)\rangle$ solution is stationary.

5.3 The Pumped Current Is the Berry Curvature

The number of particles pumped through an arbitrary cross section of the one-dimensional lattice, in the duration T of a quasi-adiabatic cycle, is evaluated combining Eqs. (5.25) and (5.42). We define the *momentum- and time-resolved current of the filled band* as

$$j_{m+1/2}^{(1)}(k, t) = \frac{1}{N} \langle \tilde{u}_1(k, t)| \, \partial_k \hat{H}(k, t) \, |\tilde{u}_1(k, t)\rangle, \qquad (5.44)$$

and perform the usual substitution $\frac{1}{N} \sum_{k \in BZ} \cdots = \int_{BZ} \frac{dk}{2\pi} \cdots$, yielding the following formula for the number of pumped particles:

$$\mathcal{Q} = \int_0^T dt \int_{BZ} \frac{dk}{2\pi} j_{m+1/2}^{(1)}(k, t). \qquad (5.45)$$

In the rest of this section, we show that the relevant contribution of the momentum- and time-resolved current is the Berry curvature associated to the filled band, and therefore the number of pumped particles is the Chern number, which in turn is indeed an integer.

To this end, we insert the result (5.42) to the definition (5.44). The contribution that incorporates two lower-band wave functions $|u_1(k, t)\rangle$, is finite; however, its integral over the Brillouin Zone vanishes, and therefore we disregard it as it does not contribute to particle pumping. Hence the leading relevant contribution is the one incorporating one filled-band $|u_1(k, t)\rangle$ and one empty-band $|u_2(k, t)\rangle$ wavefunction:

$$j_{m+1/2}^{(1)}(k, t) = i \frac{\langle u_1| \, [\partial_k \hat{H}] \, |u_2\rangle \, \langle u_2| \, \partial_t \, |u_1\rangle}{E} + c.c. \qquad (5.46)$$

where the k and t arguments are suppressed for brevity.

Now we use

$$\langle u_1| \, [\partial_k \hat{H}] \, |u_2\rangle = (E_1 - E_2) \, \langle \partial_k u_1 | u_2\rangle = -E \, \langle \partial_k u_1 | u_2\rangle, \qquad (5.47)$$

which has a straightforward proof using the spectral decomposition $\hat{H} = E_1 |u_1\rangle \langle u_1| + E_2 |u_2\rangle \langle u_2|$ of the Hamiltonian and the fact that $\partial_k \langle u_1| u_2\rangle = 0$. Therefore,

$$j_{m+1/2}^{(1)} = -i \, \langle \partial_k u_1 | u_2\rangle \, \langle u_2| \, \partial_t \, |u_1\rangle + c.c.. \qquad (5.48)$$

Since we use the parallel-transport gauge, we can replace the projector $|u_2\rangle \langle u_2|$ with unity in the preceding formula, hence the latter can be simplified as

$$j^{(1)}_{m+1/2} = -i \langle \partial_k u_1 | \partial_t u_1 \rangle + c.c. = -i (\langle \partial_k u_1 | \partial_t u_1 \rangle - \langle \partial_t u_1 | \partial_k u_1 \rangle)$$

$$= -i (\partial_k \langle u_1 | \partial_t u_1 \rangle - \partial_t \langle u_1 | \partial_k u_1 \rangle) . \qquad (5.49)$$

This testifies that the momentum- and time-resolved current is indeed the Berry curvature corresponding to the filled band, and thereby confirms the result promised in Eq. (5.3).

As a straightforward application of our result, we calculate the time dependence of the current and the number of pumped particles through an arbitrary cross section in the smoothly modulated Rice-Mele model, see Eq. (5.2). The results corresponding to four different values of the parameter $\bar{v}$ are shown in Fig. 5.2. The momentum- and time-resolved current $j^{(1)}_{m+1/2}(k, t)$ of the filled band can be obtained analytically from Eq. (5.49). Then, the time-resolved current j is defined as the integrand of the t integral in Eq. (5.45). We evaluate j via a numerical k integration, and plot the result in Fig. 5.2a. The number of pumped particles as a function of time is then evaluated numerically via $\mathcal{Q}(t) = \int_0^t dt' j(t')$; the results are shown in Fig. 5.2b. These results confirm that the number of particles pumped through the cross section during the complete cycle is an integer, and is given by the Chern number associated to the pumping cycle.

In this chapter, we have provided a formal description of adiabatic pumping in one-dimensional lattices. After identifying the current operator describing particle flow at a cross section of the lattice, we discussed the quasi-adiabatic time evolution of the lower-band states in a two-band model, and combined these results to express the number of pumped particles in the limit of adiabatic pumping. The central result is that the relevant part of the momentum- and time-resolved current carried by the lower-band electrons is the Berry curvature associated to their band.

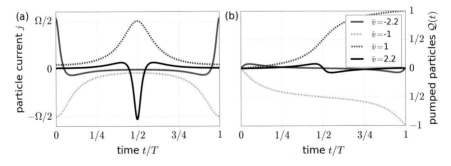

Fig. 5.2 Time dependence of the current and the number of pumped particles in an adiabatic cycle

Problems

5.1 The smooth pump sequence of the Rice-Mele model

For the smoothly modulated Rice-Mele pumping cycle, see (5.2), evaluate the momentum- and time-dependent current density, and the number of particles pumped through an arbitrary unit cell boundary as the function of time; that is, reproduce Fig. 5.2.

5.2 Parallel-transport time parametrization

Specify a parallel-transport time parametrization for the ground state of the two-level Hamiltonian defined by Eqs. (5.1), (5.2), and (a) $k = 0$ (b) $k = \pi$.

5.3 Quasi-adiabatic dynamics with a different boundary condition

In Sect. 5.2.2, we described a *stationary state* of a quasi-adiabatically driven two-level system, and used the result to express the number of particles pumped during a complete cycle. How does the derivation and the result change, if we describe the dynamics not via the stationary state, but by specifying that the initial state is the instantaneous ground state of the Hamiltonian at $t = 0$? Is the final result for the number of pumped particles different in this case?

5.4 Adiabatic pumping in multiband models

Generalize the central result of this chapter in the following sense. Consider adiabatic charge pumping in a one-dimensional multi-band system ($n = 1, 2, \ldots, N_b$), where the energies of the first N_{filled} bands ($n = 1, 2, \ldots, N_{\text{filled}}$) are below the Fermi energy and the energies of the remaining bands ($n = N_{\text{filled}} + 1, \ldots, N_b$) are above the Fermi energy, and the bands do not cross each other. Show that the number of particles adiabatically pumped through an arbitrary cross section of the crystal is the sum of the Chern numbers of the filled bands.

Chapter 6
Two-Dimensional Chern Insulators: The Qi-Wu-Zhang Model

The unique physical feature of topological insulators is the guaranteed existence of low-energy states at their boundaries. We have seen an example of this for a one-dimensional topological insulator, the SSH model: A finite, open, topologically nontrivial SSH chain hosts zero-energy bound states at both ends. The bulk–boundary correspondence was the way in which the topological invariant of the bulk—in the case of the SSH chain, the winding number of the bulk Hamiltonian—can be used to predict the number of edge states.

We will show that the connection between the Chern number and the number of edge-state channels is valid in general for two-dimensional insulators. This is the statement of bulk–boundary correspondence for Chern insulators. The way we will show this inverts the argument above: taking any two-dimensional insulator, we can map it to an adiabatic pump sequence in a one-dimensional insulator by demoting one of the wavenumbers to time. The connection between the Chern number and the number of edge states in the higher dimensional Hamiltonian is a direct consequence of the connection between Chern number and charge pumping in the lower dimensional system.

Chern insulators (two-dimensional band insulators with nonvanishing Chern number) were first used to explain the Quantum Hall Effect. There an external magnetic field, included in lattice models via a Peierls substitution, is responsible for the nonzero value of the Chern number. Peierls substitution, however, breaks the lattice translation invariance. This necessitates extra care, including the use of magnetic Brillouin zones whose size depends on the magnetic field.

The models we construct in this chapter describe the so-called Quantum Anomalous Hall Effect. Here we have the same connection between edge states and bulk Chern number as in the Quantum Hall Effect, however, there is no external magnetic field, and thus no complications with magnetic Brillouin zones. The Quantum Anomalous Hall Effect has recently been observed in thin films of chromium-doped $(Bi,Sb)_2Te_3$ [8].

© Springer International Publishing Switzerland 2016
J.K. Asbóth et al., *A Short Course on Topological Insulators*, Lecture Notes in Physics 919, DOI 10.1007/978-3-319-25607-8_6

To illustrate the concepts of Chern insulators, we will use a toy model introduced by Qi, Wu and Zhang [24], which we call the QWZ model. This model is also important because it forms the basic building block of the Bernevig-Hughes-Zhang model for the Quantum Spin Hall Effect (Chap. 8), and thus it is also sometimes called "half BHZ".

6.1 Dimensional Extension: From an Adiabatic Pump to a Chern Insulator

We want to construct a two-dimensional lattice Hamiltonian $\hat{H}$ with a nonvanishing bulk Chern number. We will do this by first constructing the bulk momentum-space Hamiltonian $\hat{H}(k_x, k_y)$, from which the real-space Hamiltonian can be obtained by Fourier transformation. For the construction we simply take an adiabatic pump sequence on a one-dimensional insulator, $\hat{H}(k, t)$, and reinterpret the cyclic time variable t as a new momentum variable k_y. This way of gaining an extra dimension by *promotion* of a cyclic parameter in a continuous ensemble to a momentum is known as *dimensional extension*. This, and the reverse process of *dimensional reduction*, are key tools to construct the general classification of topological insulators [29].

6.1.1 *From the Rice-Mele Model to the Qi-Wu-Zhang Model*

To see how the construction of a Chern insulator works, we take the example of the smooth pump sequence on the Rice-Mele model from the previous chapter, Eq. (4.7). In addition to the promotion of time to an extra wavenumber, $\Omega t \rightarrow k_y$, we also do an extra unitary rotation in the internal Hilbert space, to arrive at the Qi-Wu-Zhang model,

$$\hat{H}(k) = \sin k_x \hat{\sigma}_x + \sin k_y \hat{\sigma}_y + [u + \cos k_x + \cos k_y]\hat{\sigma}_z. \tag{6.1}$$

The mapping is summarized in Table 6.1.

Table 6.1 Mapping of an adiabatic pump sequence of the Rice-Mele model, $\hat{H}(k, t)$ to the QWZ model for the Anomalous Hall Effect, $\hat{H}(k_x, k_y)$

Adiabatic pump in the RM model	QWZ model (Chern insulator)
Average intracell hopping $\overline{v}$	Staggered onsite potential u
Wavenumber $k \in [0, 2\pi)$	Wavenumber $k_x \in [0, 2\pi)$
Time $t \in [0, T)$	Wavenumber $k_y \in [0, 2\pi)$
$\sigma_x, \sigma_y, \sigma_z$	$\sigma_y, \sigma_z, \sigma_x$

The corresponding $\mathbf{d}(\mathbf{k})$ vector reads,

$$\mathbf{d}(k_x, k_y) = \begin{pmatrix} \sin k_x \\ \sin k_y \\ u + \cos k_x + \cos k_y \end{pmatrix}. \tag{6.2}$$

6.1.2 Bulk Dispersion Relation

We can find the dispersion relation of the QWZ model using the algebraic properties of the Pauli matrices, whereby $\hat{H}^2 = E(\mathbf{k})\mathbb{I}_2$, with $\mathbb{I}_2$ the unit matrix. Thus, the spectrum of the QWZ model has two bands, the two eigenstates of $\hat{H}(\mathbf{k})$, with energies

$$E_{\pm}(k_x, k_y) = \pm|\mathbf{d}(k_x, k_y)| \tag{6.3}$$

$$= \pm\sqrt{\sin^2(k_x) + \sin^2(k_y) + (u + \cos(k_x) + \cos(k_y))^2}. \tag{6.4}$$

The spectrum of the QWZ model is depicted in Fig. 6.1.

There is an energy gap in the spectrum of the QWZ model, which closes at finetuned values of $u = +2, 0, -2$. This is simple to show, since the gap closing requires $\mathbf{d}(\mathbf{k}) = 0$ at some $\mathbf{k}$. From Eq. (6.2), $d_x(\mathbf{k}) = d_y(\mathbf{k}) = 0$ restricts us to four inequivalent points in the Brillouin zone:

- if $u = -2$: at $k_x = k_y = 0$, the so-called Γ point;
- if $u = 0$: at $k_x = 0, k_y = \pi$ and $k_x = \pi, k_y = 0$, two inequivalent so-called X points;
- if $u = +2$: at $k_x = \pi, k_y = \pi$, the so-called M point; note that $k_x = \pm\pi$, $k_y = \pm\pi$ are all equivalent

In the vicinity of a gap closing point, called *Dirac point*, the dispersion relation has the shape of a *Dirac cone*, as seen in Fig. 6.1. For all other values of $u \neq -2, 0, 2$, the spectrum is gapped, and thus it makes sense to investigate the topological properties of the system.

6.1.3 Chern Number of the QWZ Model

Although we calculated the Chern number of the corresponding pump sequence in the previous chapter, we show the graphical way to calculate the Chern number of the QWZ model. We simply count how many times the torus of the image of the Brillouin zone in the space of $\mathbf{d}$ contains the origin. To get some feeling about the not so trivial geometry of the torus, it is instructive to follow a gradual sweep of

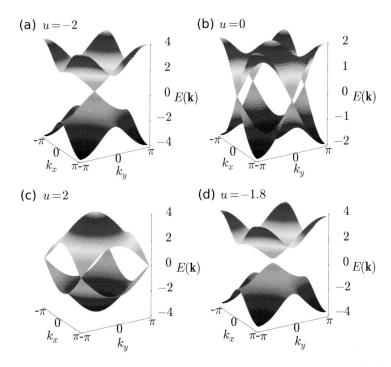

Fig. 6.1 The bulk dispersion relation of the QWZ model, for various values of u, as indicated in the plots. In (**a**)–(**c**), the gapless cases are shown, where the bulk gap closes at so-called Dirac points. In (**d**), with a generic value $u = -1.8$, the system is insulating

the Brillouin zone in Fig. 6.2. The parameter u shifts the whole torus along the d_z direction, thus as we tune it we also control whether the origin is contained inside it or not. For the QWZ model three situations can occur as depicted in Fig. 6.3. It can happen that the torus does not contain the origin, as in (a) and (d), and the Chern number is $Q = 0$. This is the case for $|u| > 2$. It can also happen that the origin is in the inside of the torus: a line from the origin to infinity will then inevitably pierce the torus. The first piercing can be from the blue side (outside) of the surface as in (b), with $Q = -1$ – for $-2 < u < 0$ –, or from the red side (inside) as in (c), with $Q = 1$ – for $0 < u < 2$.

To summarize, the Chern number Q of the QWZ model is

$$u < -2 \quad : \quad Q = 0; \tag{6.5a}$$

$$-2 < u < 0 \quad : \quad Q = -1; \tag{6.5b}$$

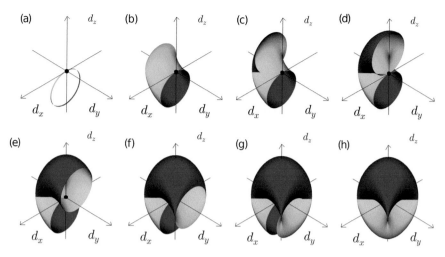

Fig. 6.2 The surface $\mathbf{d}(\mathbf{k})$ for the QWZ model as $\mathbf{k}$ sweeps through the whole Brillouin zone. To illustrate how this surface is a torus the sweeping is done gradually with $u = 0$. In (**a**) the image of the $k_y = -\pi$ line is depicted. In (**b**) the image for the region $k_y = -\pi \cdots - 0.5\pi$, in (**c**) $k_y = -\pi \cdots -0.25\pi$, in (**d**) $k_y = -\pi \cdots -0$, in (**e**) $k_y = -\pi \cdots 0.25\pi$, in (**f**) $k_y = -\pi \cdots 0.5\pi$, in (**g**) $k_y = -\pi \cdots 0.75\pi$ and finally in (**h**) the image of the whole Brillouin zone is depicted and the torus is closed

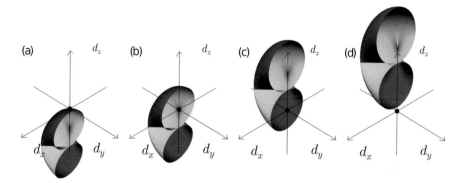

Fig. 6.3 The torus $\mathbf{d}(\mathbf{k})$ of the QWZ model for different values of u. For clarity only the image of half of the Brillouin zone is shown. In (**a**) and (**d**) $u = \mp 2.2$ and the torus does not contain the origin hence $Q = 0$. In (**b**) $u = -1$, taking an infinite line from the origin along the positive z axis we hit the blue side of the torus once hence $Q = -1$. In (**c**) $u = 1$, taking the infinite line in the negative z direction we hit the red side of the torus thus $Q = 1$

$$0 < u < 2 \quad : \quad Q = +1; \tag{6.5c}$$

$$2 < u \quad : \quad Q = 0. \tag{6.5d}$$

6.1.4 The Real-Space Hamiltonian

We obtain the full Hamiltonian of the Qi-Wu-Zhang model by inverse Fourier transform of the bulk momentum-space Hamiltonian, Eq. (6.1), as

$$
\hat{H} = \sum_{m_x=1}^{N_x-1} \sum_{m_y=1}^{N_y} \left(\left| m_x+1, m_y \right\rangle \left\langle m_x, m_y \right| \otimes \frac{\hat{\sigma}_z + i\hat{\sigma}_x}{2} + h.c. \right)
$$

$$
+ \sum_{m_x=1}^{N_x} \sum_{m_y=1}^{N_y-1} \left(\left| m_x, m_y+1 \right\rangle \left\langle m_x, m_y \right| \otimes \frac{\hat{\sigma}_z + i\hat{\sigma}_y}{2} + h.c. \right)
$$

$$
+ u \sum_{m_x=1}^{N_x} \sum_{m_y=1}^{N_y} \left| m_x, m_y \right\rangle \left\langle m_x, m_y \right| \otimes \hat{\sigma}_z. \qquad (6.6)
$$

As sketched in Fig. 6.4, the model describes a particle with two internal states hopping on a lattice where the nearest neighbour hopping is accompanied by an operation on the internal degree of freedom, and this operation is different for the hoppings along the x and y directions. In addition, there is a staggered onsite potential of strength u (if we envision the internal degree of freedom as representing two sites within the unit cell). Unlike in the case of the SSH model (Chap. 1), the real-space form of the QWZ Hamiltonian is not intuitive.

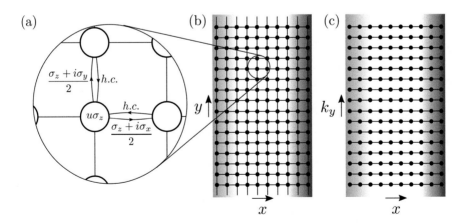

Fig. 6.4 Sketch of the QWZ model: a particle with two internal states hopping on a square lattice. (**a**): The onsite potential and the hopping amplitudes are operators acting on the internal states. (**b**): A strip, with periodic boundary conditions along y, open boundaries along x (hopping amplitudes set to zero). Light blue/dark red highlights left/right edge region. (**c**): Upon Fourier transformation along y, the strip falls apart to an ensemble of one-dimensional Hamiltonians, indexed by k_y

6.2 Edge States

We constructed a Chern insulator using an adiabatic charge pump. As we saw in Chap. 4, charge pumps also induce energy eigenstates at the edge regions that cross from negative energy to positive energy bands, or vice versa. What do these energy eigenstates correspond to for the Chern insulator?

6.2.1 Dispersion Relation of a Strip Shows the Edge States

To see edge states, consider a strip of a two-dimensional insulator depicted in Fig. 6.4. Along y, we take periodic boundary conditions (close the strip to a cylinder), and go to the limit $N_y \to \infty$. Along x, the strip is terminated by setting the hopping amplitudes to 0 (open boundary condition), and it consists of N sites. Translation invariance holds along y, so we can partially Fourier transform—only along y. After the Fourier transformation, the original Hamiltonian falls apart to a set of one-dimensional lattice Hamiltonians indexed by a continuous parameter k_y, the wavenumber along y. For the QWZ model, Eq. (6.6), the k_y-dependent Hamiltonian reads

$$\hat{H}(k_y) = \sum_{m_x=1}^{N_x-1} \left(|m_x + 1\rangle \langle m_x| \otimes \frac{\hat{\sigma}_z + i\hat{\sigma}_x}{2} + h.c. \right) +$$

$$\sum_{m_x=1}^{N_x} |m_x\rangle \langle m_x| \otimes (\cos k_y \hat{\sigma}_z + \sin k_y \hat{\sigma}_y + u\hat{\sigma}_z). \qquad (6.7)$$

Note that this is the same dimensional reduction argument as before, but for a system with edges. Energy eigenstates $|\Psi(k_y)\rangle$ of the strip fall into the categories of bulk states and edge states, much as in the one-dimensional case. All states are delocalized along y, but bulk states are also delocalized along x, while edge states are exponentially confined to the left ($x = 1$) or the right ($x = N$) edge. If we find energy eigenstates with energy deep in the bulk gap, they have to be edge states, and can be assigned to the left or the right edge.

An example for the dispersion relation of a strip is shown in Fig. 6.5, edge states on the left/right edge highlighted using light blue/dark red. We used the QWZ model, strip width $N = 10$, sublattice potential parameter $u = -1.5$, and the same practical definition of edge states as in the Rice-Mele model,

$$|\Psi(k_y)\rangle \text{ is on the right edge} \Leftrightarrow \sum_{m_x=N-1}^{N} \sum_{\alpha \in \{A,B\}} |\langle \Psi(k_y) | m_x, \alpha \rangle|^2 > 0.6; \qquad (6.8)$$

$$\left|\Psi(k_y)\right\rangle \text{ is on the left edge } \Leftrightarrow \sum_{m_x=1}^{2} \sum_{\alpha \in \{A,B\}} \left|\langle \Psi(k_y) \mid m_x, \alpha \rangle\right|^2 > 0.6. \qquad (6.9)$$

6.2.2   Edge States Conduct Unidirectionally

Notice the edge state branches of the dispersion relation of the QWZ strip, Fig. 6.5, which connect the lower and upper band across the bulk gap. They are the edge states of the pumped Rice-Mele model, Fig. 4.6, but we now look at them with a new eye. For the edge states in the QWZ model, dE/dk_y corresponds to the group velocity along the edge. Thus, the dispersion relation tells us that *particles in the QWZ model at low energy are confined either to the to left edge and propagate upwards, or to the right edge and propagate downwards.*

The presence of one-way conducting edge state branches implies that the QWZ model is no longer, strictly speaking, an insulator. Because of the bulk energy gap, it cannot conduct (at low energies) between the left and right edges. However, it will conduct along each edge.

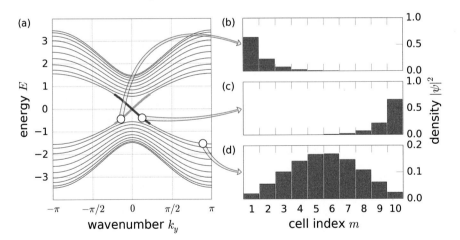

Fig. 6.5 Dispersion relation of a strip of the QWZ model, of width $N = 10$, and sublattice potential parameter $u = -1.5$. Because the strip is translation invariant along the edge, the wavenumber k_y is a good quantum number, and the energy eigenvalues can be plotted (**a**) as a function of k_y, forming branches of a dispersion relation. Light blue (dark red) highlights energies of edge states, whose wavefunction has over 60 % weight on unit cells with $m_x \leq 2$ (with $m_x \geq N - 1$). These are parts of the Nth and $N + 1$th branch, which split off from the bulk around $-\pi/4 < k_y < \pi/4$, and have an avoided crossing with an exponentially small gap near $k_y = 0$. We show the marginal position probability distribution of the Nth energy eigenstate, $P_N(m_x) = \sum_{\alpha \in \{A,B\}} \sum_{m_y} |\langle m_x, \alpha | \Psi_N \rangle|^2$, for three values of k_y. Depending on k_y, this state can be an edge state on the right edge (**b**), on the left edge (**c**), or a bulk state (**d**)

6.2.3 Edge States and Edge Perturbation

We can use dimensional reduction and translate the discussion about the robustness of edge states from Chap. 4 to the edge states of the QWZ model. This treats the case where the Hamiltonian is modified in a way that it only acts in the edge regions, and is translation invariant along the edges. As an example, we introduce an extra, state-independent next-nearest neighbor hopping, and onsite potentials at the left and right edge of the sample. As Fig. 6.6 shows, this can modify the existing edge state branches, as well as create new edge state branches by deforming bulk branches. Including the new local terms the Hamiltonian of Eq. (6.7) is augmented to read

$$\hat{H}(k_y) = \sum_{m_x=1}^{N_x-1} \left(|m_x + 1\rangle \langle m_x| \otimes \frac{\hat{\sigma}_z + i\hat{\sigma}_x}{2} + h.c. \right) +$$

$$\sum_{m_x=1}^{N_x} |m_x\rangle \langle m_x| \otimes (\cos k_y \hat{\sigma}_z + \sin k_y \hat{\sigma}_y + u\hat{\sigma}_z) +$$

$$\sum_{m_x \in \{1,N\}} |m_x\rangle \langle m_x| \otimes \hat{\mathbb{1}}_2 \left(\mu^{(m_x)} + h_2^{(m_x)} \cos 2k_y \right). \qquad (6.10)$$

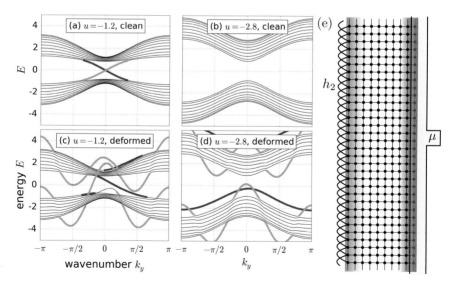

Fig. 6.6 Dispersion relation of a strip of the QWZ model, with edge states on the left/right edge highlighted in light blue/dark red. Top row: clean system. Bottom row: with extra next nearest neighbor hopping along y, on the left edge, with amplitude $h_2^{(1)} = 2$ and an onsite potential $\mu^{(N)}$ on the right edge. The value of the potential in (**c**) is $\mu^{(N)} = 0.5$ and in (**d**) it is $\mu^{(N)} = 1.5$. In (**e**) the schematics of the considered perturbations is shown. The additional potential terms do not affect the bulk states but distort the edge modes and bring in new edge modes at energies that are in the bulk gap or above/below all bulk energies

Here, $\mu^{(1)/(N)}$ is the onsite potential on the left/right edge and $h_2^{(1)/(N)}$ describes a second nearest neighbor hopping on the left/right edge. If not stated otherwise these terms are considered to be zero.

In the top row of Fig. 6.6 the spectrum of strips without edge perturbations are depicted for Chern number $Q = -1$ (a) and $Q = 0$ (b) respectively. As we expected, a nonzero Chern number results in edge states, one on each edge. In (c) and (d) switching on perturbations, we see new edge states moving in to the gap. The onsite potential acts as an overall shift in energy on the states around $m_x = N$, the second nearest neighbor hopping adds a considerable warping to the states localized around $m_x = 1$. The deformations can change the number of edge states at a specific energy, but only by adding a pair of edge states with opposite propagation directions. This leaves the topological invariant unchanged.

6.2.4   Higher Chern Numbers by Coupling Layers

A systematic way to construct models with higher Chern numbers is to layer sheets of Chern insulators onto each other, as illustrated in Fig. 6.7. The single-particle Hilbert space of the composite system of D layers is a direct sum of the Hilbert spaces of the layers.

$$\mathscr{H}_D = \mathscr{H}_{L1} \oplus \mathscr{H}_{L2} \oplus \ldots \oplus \mathscr{H}_{LD}. \tag{6.11}$$

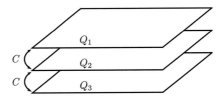

Fig. 6.7 Layering sheets of 2-dimensional insulators on top of each other is a way to construct a two-dimensional insulator with higher Chern numbers. For uncoupled layers, the Chern numbers can simply be summed to give the total Chern number of the 3-layer structure, $Q_1 + Q_2 + Q_3$. Switching on coupling (hopping terms) between the sheets cannot change the Chern number as long as the bulk gap is not closed

The Hamiltonian, including a state-independent interlayer coupling with amplitude C, is

$$\hat{H}_D = \sum_{d=1}^{D} |d\rangle \langle d| \otimes \hat{H}_{Ld} + \sum_{d=1}^{D-1} (|d+1\rangle \langle d| + |d\rangle \langle d+1|) \otimes C\,\hat{\mathbb{I}}_{2N_xN_y}, \qquad (6.12)$$

with $\hat{\mathbb{I}}_{2N_xN_y}$ the unit operator on the Hilbert space of a single layer. The operators $\hat{H}_{Ld}$ we consider below are of the form of Eq. (6.10) with different values of u, and can have an overall real prefactor. In layer d the strength of the local edge potential is denoted as $\mu^{(1)/(N)}{}_d$. As an example, the matrix of the Hamiltonian of a system with three coupled layers reads

$$H_3 = \begin{bmatrix} H_{L1} & C\mathbb{I} & 0 \\ C\mathbb{I} & H_{L2} & C\mathbb{I} \\ 0 & C\mathbb{I} & H_{L3} \end{bmatrix}. \qquad (6.13)$$

Numerical results for two and three coupled layers, with different Chern numbers in the layers, are shown in Fig. 6.8. The coupling of copropagating edge modes lifts

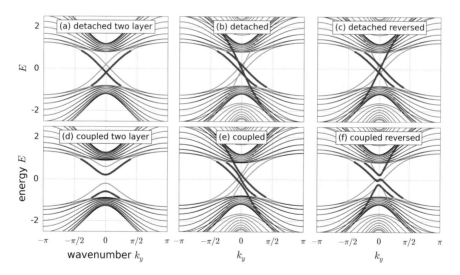

Fig. 6.8 Dispersion relations of strips of multilayered QWZ model, $N = 10$ unit cells wide. In all cases the bulk Hamiltonian of the first layer $\hat{H}_{L1}$ is the QWZ Hamiltonian with $u = -1.2$. To elucidate the interplay of the edge states we consider a finite potential acting on the edge $\mu_{L1}^{(1)/(N)} = 0.2/-0.2$. In the left column ((**a**) and (**d**)), we have two layers, with bulk $\hat{H}_{L2} = -\hat{H}_{L1}$ and edge onsite potential $\mu_{L2}^{(1)/(N)} = \mu_{L1}^{(1)/(N)}$. In the middle ((**b**), (**e**)) and right ((**c**), (**f**)) columns, we have three layers. The third layer is characterized in the bulk by $\hat{H}_{L3} = \hat{H}_{L1}$ and on the edge by $\mu_{L3}^{(1)/(N)} = -\mu_{L1}^{(1)/(N)}$. In (**b**) and (**e**) $\hat{H}_{L2} = 2\hat{H}_{L1}$, while in (**d**) and (**e**) $\hat{H}_{L2} = -2\hat{H}_{L1}$. For all four cases $\mu_{L2}^{(1)/(N)} = 0$. Coupling in (**d**), (**e**) and (**f**) is uniform with magnitude $C = 0.4$

the degeneracies, but cannot open gaps in the spectrum, except in the case of strong coupling. This is a simple consequence of the fact that an energy eigenstate has to be a single valued function of momentum. To open a gap, counterpropagating edge states have to be coupled. We achieve this by coupling layers of the QWZ model with opposite sign of the Chern number Q. For the case of two layers (Fig. 6.8, second row), this opens a gap in the spectrum. If there are three layers, there is a majority direction for the edge states, and so one edge state survives the coupling.

6.3 Robustness of Edge States

Up to now, we have considered clean edges, i.e., two-dimensional Chern insulators that were terminated by an edge (at $m_x = 1$ and $m_x = N$), but translationally invariant along the edge, along y. This translational invariance, and the resulting fact that the wavenumber k_y is a good quantum number, was used for the definition of the topological invariant $N_+ - N_-$, which was the net number of edge bands propagating along the edge, equal to the bulk Chern number, Q. With disorder in the edge region that breaks translational invariance along y, we no longer have a good quantum number k_y, and edge state bands are not straightforward to define. However, as we show in this section, the edge states must still be there in the presence of disorder, since disorder at the edges cannot close the bulk gap.

6.3.1 Smoothly Removing Disorder

Consider a finite sample of a Chern insulator, with a clean bulk part but disordered edge region, as depicted in Fig. 6.9. The bulk gap of the sample decreases due to disorder, but we suppose that it is not closed completely (just renormalized). Consider now a small part of the sample, containing some of the edge, indicated by the dotted rectangle on the right of Fig. 6.9. Although this is much smaller than the whole sample, it is big enough so that part of it can be considered as translation invariant "bulk". Now in this small part of the sample, we adiabatically (and smoothly) deform the Hamiltonian in such a way that we set the disorder gradually to 0. This includes straightening the part of the open boundary of the sample that falls into the dotted rectangle, to a straight line. The deformation is adiabatic in the sense that the bulk gap is not closed in the process. Since this small part is a clean Chern insulator, with a bulk Chern number of Q, it can be deformed in such a way that the only edge states it contains are $|Q|$ states propagating counterclockwise (if $Q > 0$, say).

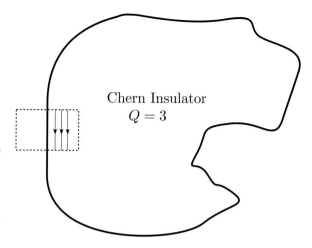

Fig. 6.9 A disordered sample of Chern insulator. The dotted lines indicate rectangular parts of the sample, where disorder can be turned off adiabatically to reveal edge states (indicated in black). Since particles cannot turn back (unidirectional, or *chiral* channels), and cannot go into the bulk (in the gap), they have to travel all the way on the perimeter of the disordered sample, coming back to the rectangular, clean part

6.3.2 Unitarity: Particles Crossing the Clean Part Have to Go Somewhere

Consider a particle in an edge state in the small clean part of the sample, with energy deep inside the bulk gap ($E \approx 0$). What can its future be, as its time evolution follows the Schrödinger equation appropriate for this closed system? Since the edge state is a chiral mode, the particle has to propagate along the edge until it leaves the clean region. Because of unitarity, the particle cannot "stop" at the edge: that would mean that at the "stopping point", the divergence of the particle current in the energy eigenstate is nonzero. In other words, the particle current that flows in the edge state has to flow somewhere. (Put differently, if the mode describing an edge state particle "stopped at the interface", two particles, initially orthogonal, following each other in the mode, would after some time arrive to the same final state. This would break unitarity.) After leaving the clean part of the sample, the particle cannot propagate into the bulk, since its energy is deep in the bulk gap. The disorder in the clean part was removed adiabatically, and thus there are no edge states at the interface of the clean part and the disordered part of the sample, along the dashed line. The particle cannot turn back, as there are no edge states running "down" along the edge in the clean part. The only thing the particle can do is propagate along the edge, doing a full loop around the sample until it comes back to the clean part from below again.

The argument of the previous paragraph shows that even though the sample is disordered, there has to be a low energy mode that conducts perfectly (reflectionless) along the edge. Since at zero energy there are Q orthogonal states a particle can be in at the edge of the clean part of the sample, unitarity of the dynamics of the particles requires that all along the edge of the disordered sample there are Q orthogonal modes that conduct counterclockwise. There can be additional low energy states, representing trapped particles, or an equal number of extra edge states for particles propagating counterclockwise and clockwise. However, the total number of counterclockwise propagating edge modes at any part of the edge always has to be larger by Q than the number of clockwise propagating edge modes. Because the Hamiltonian is short range, our conclusions regarding the number of edge states at any point far from the deformed region have to hold independent of the deformation.

To be precise, in the argument above we have shown the existence of Q edge states all along the edge of the sample, except for the small part that was adiabatically cleaned from disorder. One way to finish the argument is by considering another part of the sample. If we now remove the disorder adiabatically only in this part, we obtain the existence of the edge modes in parts of the sample including the original dotted rectangle, which was not covered by the argument of the previous paragraph.

Problems

6.1 Phase diagram of the anisotropic QWZ model

The lattice Hamiltonian of the QWZ model is provided in Eq. (6.6). Consider the anisotropic modification of the Hamiltonian when the first term of Eq. (6.6), describing the hopping along the x axis, is multiplied by a real number A. Plot the phase diagram of this model, that is, evaluate the Chern number as a function of two parameters u and A.

Chapter 7
Continuum Model of Localized States at a Domain Wall

So far, we have discussed edge states in lattice models, in which the states live on discrete lattice sites, and the Hamiltonian governing the physics is a matrix. In this chapter, we argue that in certain cases, it is also possible to describe these states via a continuum model, in which the states live in continuous space, and the Hamiltonian is a differential operator. One benefit of such a continuum description is that it allows one to use the vast available toolkit of differential equations for solid-state problems in general, including the description of topologically protected states in particular. Another interesting aspect of these continuum models is their strong similarity with the Dirac equation describing relativistic fermions. A limitation of the continuum models is that their validity is restricted to narrow windows in momentum and energy; typically they are applied in the vicinities of band edges. Here, we obtain the continuum differential equations for three basic lattice models: the one-dimensional monatomic chain, the one-dimensional SSH model, and the two-dimensional QWZ model. In the cases of the SSH and QWZ models, the resulting equations will be used to analytically characterize the localized states appearing at boundaries between regions with different topological invariants. Even though in the entire chapter we build our discussion on the three specific lattice models, the applicability of the technique introduced here, called *envelope-function approximation*, is more general and widely used to describe electronic states in various crystalline solids.

7.1 One-Dimensional Monatomic Chain in an Electric Potential

We use this minimal lattice model to illustrate the basic concepts of the envelope-function approximation (EFA) [3], the technique that allows us to find the continuum versions of our lattice models. Of course, the one-dimensional monatomic chain does not host topologically protected states.

© Springer International Publishing Switzerland 2016

J.K. Asbóth et al., *A Short Course on Topological Insulators*, Lecture Notes in Physics 919, DOI 10.1007/978-3-319-25607-8_7

7.1.1   The Model

We take a long lattice with $N \gg 1$ unit cells (or sites) without an internal degree of freedom, with periodic boundary conditions, and a negative hopping amplitude $t < 0$. We consider the situation when the electrons are subject to an inhomogeneous electric potential $V(x)$. This setup is pictured in Fig. 7.1. The lattice Hamiltonian describing this inhomogeneous system reads

$$H_{\rm i} = H + V, \tag{7.1}$$

where

$$H = t \sum_{m=1}^{N} |m\rangle \langle m+1| + h.c., \tag{7.2}$$

$$V = \sum_{m=1}^{N} V_m |m\rangle \langle m|, \tag{7.3}$$

with $V_m = V(x = m)$. Our aim is to construct a continuum model that accurately describes the low-energy eigenstates of this lattice Hamiltonian.

Before discussing the inhomogeneous case incorporating $V(x)$, focus first on the homogeneous system. The bulk momentum-space Hamiltonian is a scalar (1×1 matrix) in this model, since there is no integral degree of freedom associated to the unit cell; it reads

$$H(k) = \epsilon(k) |k\rangle \langle k|, \tag{7.4}$$

where $\epsilon(k)$ is the electronic dispersion relation:

$$\epsilon(k) = -2|t| \cos k. \tag{7.5}$$

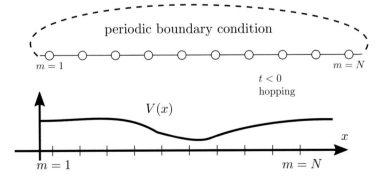

Fig. 7.1 One-dimensional monatomic chain in an inhomogeneous potential $V(x)$

The low-energy part of the dispersion relation is located around zero momentum, and is approximated by a parabola:

$$\epsilon(k) \approx -2|t| + |t|k^2 = \epsilon_0 + \frac{k^2}{2m^*}, \tag{7.6}$$

where we introduced the minimum energy of the band $\epsilon_0 = -2|t|$, and the effective mass $m^* = \frac{1}{2|t|}$ characterizing the low-energy part of the dispersion relation. (Using proper physical units, the effective mass would have the form $m^* = \hbar^2/(2|t|a^2)$, with a being the lattice constant.) For simplicity, we suppress ϵ_0 in what follows; that is, we measure energies with respect to ϵ_0.

7.1.2   Envelope-Function Approximation

Our goal is to find the low-energy eigenstates of the inhomogeneous lattice Hamiltonian H_i. The central proposition of the EFA, applied to our specific example of the one-dimensional monatomic chain, says that it is possible to complete this goal by solving the simple continuum Schrödinger equation

$$H_{\mathrm{EFA}}\varphi(x) = E\varphi(x), \tag{7.7}$$

where the *envelope-function Hamiltonian* H_{EFA} has a very similar form to the free-electron Hamiltonian with the electric potential V:

$$H_{\mathrm{EFA}} = \frac{\hat{p}^2}{2m^*} + V(x). \tag{7.8}$$

Here $\hat{p} = -i\partial_x$ is the usual real-space representation of the momentum operator, and the function $\varphi(x)$ is usually called *envelope function*. Note the very simple relation between the low-energy dispersion in Eq. (7.6) and the kinetic term in the EFA Hamiltonian (7.8): the latter can be obtained from the former by substituting the momentum operator $\hat{p}$ in place of the momentum k.

Before formulating the EFA proposition more precisely, we introduce the concept of a *spatially slowly varying* envelope function. We say that $\varphi(x)$ is spatially slowly varying, if its Fourier transform

$$\tilde{\varphi}(q) = \int_0^N dx \frac{e^{-iqx}}{\sqrt{N}}\varphi(x) \tag{7.9}$$

is localized to the $|q| \ll \pi$ region, i.e., to the vicinity of the center of the Brillouin zone.

With this definition at hand, we can formulate the EFA proposition. Consider the inhomogeneous one-dimensional monatomic chain described by H_i. Assume that $\varphi(x)$ is a spatially slowly varying eigenfunction of H_{EFA} with eigenvalue E. Then, the state $|\psi\rangle$ defined on the lattice via

$$|\psi\rangle = \sum_{m=1}^{N} \varphi(x = m) |m\rangle. \tag{7.10}$$

is approximately an eigenstate of the lattice Hamiltonian H_i with eigenvalue E. The proof follows below in Sect. 7.1.3.

Note that if the envelope function $\varphi(x)$ fulfils the normalization condition

$$\int_{0}^{N} dx |\varphi(x)|^2 = 1, \tag{7.11}$$

then, for a long lattice $N \gg 1$, the lattice state $|\psi\rangle$ will also be normalized to a good accuracy:

$$\langle \psi | \psi \rangle \approx 1. \tag{7.12}$$

It is important to point out two possible interpretations of the lattice state $|\psi\rangle$ introduced in Eq. (7.10). (1) The state $|\psi\rangle$ can be interpreted as the zero-momentum band-edge eigenstate $|k = 0\rangle = \frac{1}{\sqrt{N}} \sum_{m=1}^{N} |m\rangle$ of the homogeneous system, modulated by the envelope function $\varphi(x)$ restricted to the lattice-site positions $x = m$. (2) The state $|\psi\rangle$ can also be interpreted as a wave packet, composed of those eigenstates $|k\rangle$ of the homogeneous lattice Hamiltonian H that have wave numbers k close to the band-edge wave number, the latter being zero in this case. To see this, we first Fourier-decompose $\varphi(x)$:

$$\varphi(x) = \sum_{k \in BZ} \tilde{\varphi}(k) \frac{e^{ikx}}{\sqrt{N}} \approx \sum_{k}' \tilde{\varphi}(k) \frac{e^{ikx}}{\sqrt{N}}. \tag{7.13}$$

In the approximate equality, we used the fact that $\varphi(x)$ is spatially slowly varying, i.e., its Fourier transform $\tilde{\varphi}(k)$ is localized to the central part of the Brillouin zone, and introduced the notation $\sum_{k}'$ for a wave-number sum that goes only for the central part of the Brillouin zone. By inserting Eq. (7.13) to Eq. (7.10), we find

$$|\psi\rangle = \sum_{m=1}^{N} \left(\sum_{k}' \tilde{\varphi}(k) \frac{e^{ikm}}{\sqrt{N}} \right) |m\rangle = \sum_{k}' \tilde{\varphi}(k) |k\rangle. \tag{7.14}$$

That is, $|\psi\rangle$ is indeed a packet of plane waves with small wave numbers.

7.1.3   Envelope-Function Approximation: The Proof

To prove the EFA proposition, we calculate $H_i \, |\psi\rangle$ and utilize the spatially-slowly-varying condition on $\varphi(x)$. Start with the contribution of the bulk Hamiltonian $\hat{H}$: utilizing Eq. (7.14), we find

$$H \, |\psi\rangle = \left[\sum_{k \in \mathrm{BZ}} \epsilon(k) \, |k\rangle \, \langle k| \right] \left[\sum_{q}' \tilde{\varphi}(q) \, |q\rangle \right]. \qquad (7.15)$$

Performing the scalar product yields

$$H \, |\psi\rangle = \sum_{q}' \epsilon(q) \tilde{\varphi}(q) \, |q\rangle. \qquad (7.16)$$

Using the fact that the q sum goes for the central part of the Brillouin zone, where the dispersion relation $\epsilon(q)$ is well approximated by a parabola, we find

$$H \, |\psi\rangle \approx \sum_{q}' \frac{q^2}{2m^*} \tilde{\varphi}(q) \, |q\rangle. \qquad (7.17)$$

Utilizing the definition of the plane wave $|q\rangle$, we obtain

$$H \, |\psi\rangle \approx \sum_{q}' \frac{q^2}{2m^*} \tilde{\varphi}(q) \sum_{m=1}^{N} \frac{e^{iqm}}{\sqrt{N}} \, |m\rangle, \qquad (7.18)$$

which can be rewritten as

$$H \, |\psi\rangle \approx \sum_{m=1}^{N} \left[\sum_{q}' \frac{q^2}{2m^*} \tilde{\varphi}(q) \frac{e^{iqx}}{\sqrt{N}} \right]_{x=m} |m\rangle \qquad (7.19)$$

$$= \sum_{m=1}^{N} \left[-\frac{1}{2m^*} \partial_x^2 \sum_{q}' \tilde{\varphi}(q) \frac{e^{iqx}}{\sqrt{N}} \right]_{x=m} |m\rangle \qquad (7.20)$$

$$= \sum_{m=1}^{N} \left[\frac{\hat{p}^2}{2m^*} \varphi(x) \right]_{x=m} |m\rangle. \qquad (7.21)$$

Continue with the contribution of the potential V. Using Eqs. (7.3) and (7.10), we find

$$V \left| \psi \right\rangle = \left[\sum_{m=1}^{N} V_m \left| m \right\rangle \left\langle m \right| \right] \left[\sum_{m'=1}^{N} \varphi(m') \left| m' \right\rangle \right] = \sum_{m=1}^{N} V_m \varphi(m) \left| m \right\rangle$$

$$= \sum_{m=1}^{N} [V(x)\varphi(x)]_{x=m} \left| m \right\rangle . \tag{7.22}$$

Summing up the contribution (7.21) of H and the contribution (7.22) of V, we find

$$(H + V) \left| \psi \right\rangle \approx \sum_{m=1}^{N} [H_{\mathrm{EFA}} \varphi(x)]_{x=m} \left| m \right\rangle = \sum_{m=1}^{N} [E \varphi(x)]_{x=m} \left| m \right\rangle \tag{7.23}$$

$$= E \sum_{m=1}^{N} \varphi(m) \left| m \right\rangle = E \left| \psi \right\rangle , \tag{7.24}$$

which concludes the proof.

7.2   The SSH Model and the One-Dimensional Dirac Equation

To illustrate how the EFA captures topologically protected bound states in one-dimensional, we use the SSH model described in detail in Chap. 1. The model is visualized in Fig. 1.1. The bulk Hamiltonian is characterized by two parameters, the intra-cell and inter-cell hopping amplitudes $v, w > 0$, respectively. The bulk lattice Hamiltonian of the SSH model is given in Eq. (1.1), whereas the bulk momentum-space Hamiltonian is given in Eq. (1.14). We have seen that the (v, w) parameter space is separated to two adiabatically connected partitions by the $v = w$ line. In Sect. 1.5.5, we have also seen that localized zero-energy states appear at a domain wall between two half-infinite homogeneous regions, if the two regions have different bulk topological invariants; that is, if the sign of $v - w$ is different at the two sides of the domain wall.

This is the phenomenon that we address in this section: we show that an analytical description of such localized states can be given using the EFA. First, we discuss the electronic dispersion relation of the metallic ($v = w$) and nearly metallic ($|v - w| \ll |v + w|$) homogeneous SSH model. Second, we obtain the Dirac-type differential equation providing a continuum description for the inhomogeneous SSH model (see Fig. 7.2) for the energy range in the vicinity of the bulk band gap. Finally, we solve that differential equation to find the localized zero-energy states at a domain wall. Remarkably, the analytical treatment remains useful even if the spatial structure of the domain wall is rather irregular.

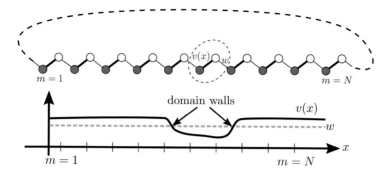

Fig. 7.2 Inhomogeneous intracell hopping and domain walls in the SSH model. The dashed ellipse denotes the unit cell. The dashed line connecting the edges of the chain denotes the periodic boundary condition

7.2.1 The Metallic Case

First, consider the metallic homogeneous SSH model, where $v = w$. The dispersion relation is shown as the blue solid line in Fig. 1.2c. The filled and empty bands touch at the end of the Brillouin zone, at $k = k_0 \equiv \pi$. Figure 1.2c shows that in the vicinity of that touching point, commonly referred to as a *Dirac point*, the dispersion relations are linear functions of the relative wave vector $q = k - k_0$. The slope of these linear functions, corresponding to the group velocity of the electrons, can be determined, e.g. by Taylor-expanding the bulk momentum-space Hamiltonian $\hat{H}(k) = (v + w \cos k)\hat{\sigma}_x + w \sin k\hat{\sigma}_y$, see Eq. (1.10), to first order in q:

$$\hat{H}(k_0 + q) \approx -wq\hat{\sigma}_y, \quad (v = w). \tag{7.25}$$

which indeed has a linear dispersion relation,

$$E_{\pm}(q) = \pm wq. \tag{7.26}$$

The eigenstates of the linearized Hamiltonian (7.25) are

$$\psi_{\pm}(q) = \frac{1}{\sqrt{2}}\begin{pmatrix} 1 \\ \mp i \end{pmatrix}. \tag{7.27}$$

Note that the dispersion relation of the Dirac equation of fermions with zero mass is

$$E_{\pm}(k) = \pm \hbar k c, \tag{7.28}$$

where $\hbar$ is the reduced Planck's constant and c is the speed of light. Comparing Eqs. (7.26) and (7.28), we conclude that the dispersion of the metallic SSH model is analogous to that of massless Dirac fermions, and the hopping amplitude of the metallic SSH model plays the role of $\hbar c$. Because of the similarity of the dispersions (7.26) and (7.28), the linearized Hamiltonian (7.25) is often called a *massless Dirac Hamiltonian*.

At this point, the linearization of the bulk momentum-space Hamiltonian of the SSH model does not seem to be a particularly fruitful simplification: to obtain the dispersion relation and the eigenstates, a 2×2 matrix has to be diagonalized, no matter if the linearization has been done or not. However, linearizing the Hamiltonian is the first step towards the EFA, as discussed below.

7.2.2   The Nearly Metallic Case

Now consider a homogeneous, insulating SSH model that is *nearly metallic*; that is, the scale of the energy gap $|v - w|$ opened at k_0 is significantly smaller than the scale of the band width $v + w$. An example is are shown in Fig. 1.2b, where the dispersion relation is plotted for the parameter values $v = 1$ and $w = 0.6$.

We wish to describe the states close to the band gap located around zero energy. Hence, again, we can use the approximate bulk momentum-space Hamiltonian obtained via linearization in the relative momentum q:

$$\hat{H}(k_0 + q) \approx M\hat{\sigma}_x - wq\hat{\sigma}_y, \tag{7.29}$$

where we defined $M = v - w$. The dispersion relation reads

$$E_\pm(q) = \pm\sqrt{M^2 + w^2 q^2}. \tag{7.30}$$

The (unnormalized) eigenstates of the linearized Hamiltonian (7.29) have the form

$$\psi_\pm(q) = \begin{pmatrix} M + iwq \\ E_\pm(q) \end{pmatrix}. \tag{7.31}$$

Note that the dispersion relation of the Dirac equation for fermions with finite mass $\mu \neq 0$ reads

$$E_\pm(k) = \pm\sqrt{\mu^2 c^4 + \hbar^2 k^2 c^2}. \tag{7.32}$$

Therefore, the parameter $M = v - w$ of the SSH model plays the role of the mass-related term μc^2 of the relativistic dispersion relation (7.32), and the linearized Hamiltonian (7.29) is often called a *massive Dirac Hamiltonian*.

7.2.3 Continuum Description of the Nearly Metallic Case

We are mostly interested in a continuum description of the zero-energy localized states formed at a domain wall between two topologically distinct regions. For simplicity and concreteness, consider the case when the domain wall is created so that the intra-cell hopping amplitude v varies in space while the inter-cell one is constant, as shown in Fig. 7.2.

In what follows, we will focus on one of the two domain walls shown in Fig. 7.2. The inhomogeneous Hamiltonian has the form

$$H_i = \sum_{m=1}^{N} v_m \left(|m, B\rangle \langle m, A| + h.c. \right) + w \sum_{m=1}^{N} \left(|m, B\rangle \langle m + 1, A| + h.c. \right),$$

(7.33)

where $v_m = v(x = m)$ and $v(x) \geq 0$ is a continuously varying function of position, which takes the constant value v_- (v_+) far on the left (right) from the domain wall.

We also assume that the local Hamiltonian is a nearly metallic SSH Hamiltonian everywhere in space. That is, $|v(x) - w| \ll v(x) + w$. This ensures that the local band gaps $|v_\pm - w|$ on the two sides of the domain wall are much smaller than the local band widths $v_\pm + w$.

Based on our experience with the EFA in the inhomogeneous one-dimensional monatomic chain (see Sect. 7.1), now we construct the EFA proposition corresponding to this inhomogeneous, nearly metallic SSH model. Recall that in the former case, we obtained the EFA Hamiltonian by (i) Taylor-expanding the bulk momentum-space Hamiltonian around the wave vector corresponding to the band extremum (that was $k_0 = 0$ in Sect. 7.1), (ii) replacing the relative wave vector q with the momentum operator $\hat{p} = -i\partial_x$, and (iii) incorporating the inhomogeneity of the respective parameter, which was the on-site potential $V(x)$ in that case. The same procedure, applied now for the SSH model with a first-order Taylor expansion, yields the following EFA Hamiltonian:

$$H_{\text{EFA}} = M(x)\hat{\sigma}_x - w\hat{p}\hat{\sigma}_y.$$

(7.34)

The EFA proposition is then formulated as follows. Assume that $\varphi(x) = (\varphi_A(x), \varphi_B(x))$ is a spatially slowly varying eigenfunction of H_{EFA} in Eq. (7.34), with eigenvalue E. Then, the state $|\psi\rangle$ defined on the lattice via

$$|\psi\rangle = \sum_{m=1}^{N} \sum_{\alpha=A,B} \varphi_\alpha(m) e^{ik_0 m} |m, \alpha\rangle$$

(7.35)

is approximately an eigenstate of the lattice Hamiltonian H_i with energy E.

Note that, in analogy with Eq. (7.14), the lattice state $|\psi\rangle$ can be reformulated as a wave packet formed by plane waves from the vicinity of the band-edge momentum k_0:

$$|\psi\rangle \approx \sum_{q}{}' \sum_{\alpha=A,B} \tilde{\varphi}_\alpha(q) \, |k_0 + q\rangle \otimes |\alpha\rangle . \qquad (7.36)$$

7.2.4 Localized States at a Domain Wall

Having the envelope-function Schrödinger equation

$$\left[M(x)\hat{\sigma}_x - w\hat{p}\hat{\sigma}_y\right]\varphi(x) = E\varphi(x) \qquad (7.37)$$

at hand, we can study the domain wall between the two topologically distinct regions. First, we consider a step-type domain wall, defined via

$$M(x) = \begin{cases} M_0 & \text{if } x > 0, \\ -M_0 & \text{if } x < 0 \end{cases} , \qquad (7.38)$$

and $M_0 > 0$, as shown in Fig. 7.3a.

We wish to use the EFA Schrödinger equation (7.37) to establish the zero-energy states localized to the domain wall, which were revealed earlier in the lattice SSH model. That is, we look for evanescent solutions of Eq. (7.37) on both sides of the domain wall, and try to match them at the domain wall at $x = 0$. For the $x > 0$ region, our evanescent-wave Ansatz reads

$$\varphi_{x>0}(x) = \begin{pmatrix} a \\ b \end{pmatrix} e^{-\kappa x} \qquad (7.39)$$

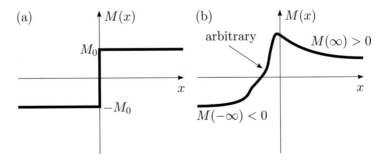

Fig. 7.3 (a) Step-like and (b) irregular spatial dependence of the mass parameter $M(x)$ of the one-dimensional Dirac equation

with $\kappa > 0$. Substituting this to Eq. (7.37) yields a quadratic characteristic equation for the energy E, having two solutions

$$E_\pm = \pm\sqrt{M_0^2 - w^2\kappa^2}. \tag{7.40}$$

The corresponding unnormalized spinors read

$$\begin{pmatrix} a_\pm \\ b_\pm \end{pmatrix} = \begin{pmatrix} \frac{M_0 - w\kappa}{E_\pm} \\ 1 \end{pmatrix}. \tag{7.41}$$

An analogous Ansatz for the $x < 0$ region is

$$\varphi_{x<0}(x) = \begin{pmatrix} c \\ d \end{pmatrix} e^{\kappa x} \tag{7.42}$$

with $\kappa > 0$, yielding the same energies as in Eq. (7.40), and the spinors

$$\begin{pmatrix} c_\pm \\ d_\pm \end{pmatrix} = \begin{pmatrix} \frac{-M_0 + w\kappa}{E_\pm} \\ 1 \end{pmatrix}. \tag{7.43}$$

Now consider an energy eigenstate with a given energy E. For clarity, set $M_0 > E \geq 0$. The (unnormalized) envelope function of the energy eigenstate has the form

$$\varphi(x) = \varphi_{x>0}(x)\Theta(x) + C\varphi_{x<0}(x)\Theta(-x), \tag{7.44}$$

where $\varphi_{x<0}$ and $\varphi_{x>0}$ should be evaluated by replacing $E_\pm \mapsto E$ and $\kappa \mapsto \frac{\sqrt{M_0^2 - E^2}}{w}$, and C is a yet unknown parameter to be determined from the boundary conditions at the domain wall. The envelope function (7.44) is an eigenstate of the EFA Hamiltonian with energy E if the boundary condition that the wave function is continuous at $x = 0$, that is,

$$\varphi_{x<0}(0) = C\varphi_{x>0}(0), \tag{7.45}$$

is fulfilled. Note that in our case, the Dirac equation is a first-order differential equation and therefore there is no boundary condition imposed on the derivative of the wave function. From the second component of Eq. (7.45), we have $C = 1$. From the first component, we have $M_0 - w\kappa = -M_0 + w\kappa$, implying $M_0 = w\kappa$ and thereby $E = 0$. The same result is obtained if the range $-M_0 < E \leq 0$ of negative energies is considered. Hence we conclude that the zero-energy state at the domain wall does appear in the continuum model of the inhomogeneous SSH chain, as expected.

Let us also determine the coefficients a and c describing this localized state:

$$a = \lim_{E \to 0} \frac{M_0 - \sqrt{M_0^2 - E^2}}{E} = 0, \tag{7.46}$$

and similarly, $c = 0$. These imply that the localized state is completely sublattice-polarized, i.e., it lives on the B sublattice, and therefore it is its own chiral partner. Considering a similar mass profile with negative M_0, we would have found that the localized state lives on the A sublattice. These properties are in line with our expectations drawn from the lattice SSH model.

A further characteristic property of the localized state is its localization length; from our continuum model, we have an analytical result for that:

$$\frac{1}{\kappa} = \frac{w}{M_0}. \tag{7.47}$$

(In physical units, that is $\frac{1}{\kappa} = \frac{w}{M_0}a$.) Recall that we are constrained to the nearly metallic regime $w \gg M_0 = v_+ - w$; together with Eq. (7.47), this implies that the localization length is much larger than one (that is, the lattice constant). This is reassuring: it means that the envelope function $\varphi(x)$ is spatially slowly varying, hence is within the range of validity of the EFA.

The result (7.47) can be compared to the corresponding result for the SSH lattice model. Equation (1.50) provides the localization length ξ of an edge state in a disordered SSH model, which corresponds to the localization length $\frac{1}{\kappa}$ obtained above. Taking the disorder-free special case of Eq. (1.50), we have

$$\xi = \frac{1}{\log \frac{w}{v}} = \frac{1}{\log \frac{w}{w+(v-w)}} \approx \frac{w}{w - v}. \tag{7.48}$$

As we are making a comparison to the nearly metallic $v \approx w$ case considered in this section, we could approximate ξ in Eq. (7.48) using a leading-order Taylor expansion in the small quantity $(w - v)/w$. The approximate result (7.48) is in line with Eq. (7.47) obtained from the continuum model.

A further interesting fact is that the existence of the localized state is not constrained to the case of a sharp, step-like domain wall described above. The simple spinor structure found above also generalizes for less regular domain walls. To see this, consider an almost arbitrary one-dimensional spatial dependence $M(x)$ of the mass, illustrated in Fig. 7.3b, with the only condition that M changes sign between the half-planes $x < 0$ and $x > 0$, i.e., $M(x \to -\infty) < 0$ and $M(x \to \infty) > 0$. We claim that there exists a zero-energy solution of the corresponding one-dimensional Dirac equation that is localized to the domain wall and has the envelope function

$$\varphi(x) = \begin{pmatrix} 0 \\ 1 \end{pmatrix} f(x). \tag{7.49}$$

To prove this claim, insert this wave function $\varphi(x)$ to the one-dimensional Dirac equation and substitute $E = 0$ therein. This procedure results in the single differential equation $\partial_x f(x) = -\frac{M(x)}{w}$, implying that Eq. (7.49) is indeed a zero-energy eigenstate of the envelope-function Hamiltonian if the function f has the form

$$f(x) = \text{const} \times e^{-\frac{1}{w}\int_0^x dx' M(x')}. \tag{7.50}$$

Furthermore, the asymptotic conditions of the mass $M(x)$ guarantee that this envelope function decays as $x \to \pm\infty$, and therefore is localized at the domain wall.

7.3   The QWZ Model and the Two-Dimensional Dirac Equation

We have introduced the QWZ model as an example for a two-dimensional Chern insulator in Chap. 6. The lattice Hamiltonian of the model is given in Eq. (6.6), whereas the bulk momentum-space Hamiltonian is given in Eq. (6.1). The dispersion relation is calculated in Eq. (6.4), and examples of it are shown in Fig. 6.1.

Recall that the model has a single parameter u, and the Chern number of the model is determined by the value of u via Eq. (6.5). Similarly to the case of the SSH model in one dimension, one can consider a domain wall between locally homogeneous regions of the QWZ model that have different Chern numbers. Just as the edge of a strip, such a domain wall can support topologically protected states that propagate along the domain wall but are localized at the domain wall in the transverse direction. The number and propagation direction of those states is determined by the magnitude and sign of the difference of the Chern numbers in the two domains, respectively. In this section, we use the EFA to provide a continuum description of such states.

7.3.1   The Metallic Case

First, consider the metallic cases of the QWZ model; that is, when the band structure has no energy gap. In particular, we will focus on the $u = -2$ case. The corresponding band structure is shown in Fig. 6.1a. The two bands touch at $\mathbf{k} = (0, 0)$, and form a Dirac cone at that Dirac point.

To describe excitations in the vicinity of the Dirac point of such a metal, it is sufficient to use a linearized approximation of the QWZ Hamiltonian $\hat{H}(\mathbf{k})$ that is obtained via a Taylor expansion of $\hat{H}(\mathbf{k})$ up to first order in the k-space location $\mathbf{q} = \mathbf{k} - \mathbf{k}_0$ measured from the Dirac point $\mathbf{k}_0$. In the case $u = -2$, the Dirac point is $\mathbf{k}_0 = (0, 0)$, and the linearized Hamiltonian reads

$$\hat{H}(\mathbf{k}_0 + \mathbf{q}) \approx q_x \hat{\sigma}_x + q_y \hat{\sigma}_y. \tag{7.51}$$

The dispersion relation is $E_\pm(\mathbf{q}) = \pm q$; again, this is analogous to that of the massless Dirac equation Eq. (7.28).

7.3.2 The Nearly Metallic Case

Now consider a QWZ insulator that is nearly metallic: $u \approx -2$. The dispersion relation for $u = -1.8$ is shown in Fig. 6.1d. In the vicinity of the metallic state, as seen in the figure, a small gap opens in the band structure at the Dirac point $\mathbf{k}_0 = (0, 0)$.

The states and the band structure around $\mathbf{k}_0$ can again be described by a linearized approximation of the QWZ Hamiltonian $\hat{H}(\mathbf{k}_0 + \mathbf{q})$ in $\mathbf{q}$:

$$\hat{H}(\mathbf{k}_0 + \mathbf{q}) \approx M \hat{\sigma}_z + q_x \hat{\sigma}_x + q_y \hat{\sigma}_y, \tag{7.52}$$

where we defined the parameter $M = u + 2$. The dispersion relation reads

$$E_\pm(\mathbf{q}) = \pm \sqrt{M^2 + q^2}. \tag{7.53}$$

A comparison with the relativistic dispersion relation (7.32) reveals that the parameter M of the QWZ model plays the role of μc^2; hence M can be called the *mass parameter*.

7.3.3 Continuum Description of the Nearly Metallic Case

We have discussed that the QWZ lattice with an inhomogeneous u parameter might support topologically protected states at boundaries separating locally homogeneous regions with different Chern numbers. Similarly to the one-dimensional SSH model treated above, these localized states can be described analytically, using the envelope function approximation (EFA), also in the two-dimensional QWZ model. In the rest of this chapter, we focus on the nearly metallic case where the inhomogeneous $u(x, y)$ is in the vicinity of -2 (i.e., $|M(x, y)| = |u(x, y) + 2| \ll 1$), in which case the low-energy excitations are expected to localize in Fourier space around the band extremum point $\mathbf{k}_0 = (0, 0)$ (see Fig. 6.1a and d). Here we obtain the EFA

Schrödinger-type equation, which resembles the two-dimensional Dirac equation, and in the next subsection we provide its localized solutions for simple domain-wall arrangements.

The considered lattice is inhomogeneous due to the spatial dependence of the parameter $u(x, y)$. In the tight-binding lattice model, we denote the value of u in unit cell $\mathbf{m} = (m_x, m_y)$ as $u_{\mathbf{m}} = u(x = m_x, y = m_y)$, and correspondingly, we introduce the *local mass parameter* via $M(x, y) = u(x, y) + 2$ and $M_{\mathbf{m}} = M(x = m_x, y = m_y)$.

The EFA Hamiltonian can be constructed the same way as in Sects. 7.1 and 7.2.3. The bulk momentum-space Hamiltonian $H(\mathbf{k}_0 + \mathbf{q})$ is Taylor-expanded around the band-edge wave vector $\mathbf{k}_0 = (0, 0)$, the wave-number components q_x and q_y are replaced by the differential operators $\hat{p}_x$ and $\hat{p}_y$, respectively, and the inhomogeneous mass parameter $M(x, y)$ is incorporated. This yields the following result in our present case:

$$\hat{H}_{\text{EFA}} = M(x, y)\hat{\sigma}_z + \hat{p}_x\hat{\sigma}_x + \hat{p}_y\hat{\sigma}_y. \tag{7.54}$$

Then, the familiar EFA proposition is as follows. Assume that the two-component envelope function $\varphi(x, y)$ is a spatially slowly varying solution of the EFA Schrödinger equation

$$\hat{H}_{\text{EFA}}\varphi(x, y) = E\varphi(x, y). \tag{7.55}$$

Then, the lattice state $|\psi\rangle$ associated to the envelope function $\varphi(x, y)$ is defined as

$$|\psi\rangle = \sum_{\mathbf{m}, \alpha} \varphi_\alpha(\mathbf{m}) |\mathbf{m}, \alpha\rangle. \tag{7.56}$$

It is claimed that the lattice state $|\psi\rangle$ is approximately an eigenstate of the inhomogeneous lattice Hamiltonian with the eigenvalue E.

7.3.4   Chiral States at a Domain Wall

We can now use the EFA Hamiltonian (7.54) to describe the chiral states at a domain wall between two topologically distinct regions of the QWZ model. The Dirac-type EFA Schrödinger equation reads:

$$\left[M(x, y)\hat{\sigma}_z + \hat{p}_x\hat{\sigma}_x + \hat{p}\hat{\sigma}_y\right]\varphi(x, y) = E\varphi(x, y). \tag{7.57}$$

Consider the homogeneous case first: $M(x, y) = M_0$, where M_0 might be positive or negative. What is the dispersion relation for propagating waves? What are the energy eigenstates? The answers follow from the plane-wave Ansatz

$$\varphi(x, y) = \begin{pmatrix} a \\ b \end{pmatrix} e^{iq_x x} e^{iq_y y} \tag{7.58}$$

with $q_x, q_y \in \mathbb{R}$ and $a, b \in \mathbb{C}$. With this trial wave function, Eq. (7.57) yields two solutions:

$$E_\pm = \pm\sqrt{M_0^2 + q_x^2 + q_y^2}, \tag{7.59}$$

and

$$\frac{a_\pm}{b_\pm} = \frac{q_x - iq_y}{E_\pm - M_0}. \tag{7.60}$$

Describe now the states at a domain wall between two locally homogeneous regions where the sign of the mass parameter is different. Remember that the sign of the mass parameter in the EFA Hamiltonian is related to the Chern number of the corresponding homogeneous half-BHZ lattice: in our case, a positive (negative) mass implies a Chern number -1 (0).

To be specific, we will consider the case when the two domains are defined as the $y < 0$ and the $y > 0$ half-planes, i.e., the mass profile in Eq. (7.57) are

$$M(x, y) = \begin{cases} M_0 & \text{if } y > 0, \\ -M_0 & \text{if } y < 0 \end{cases} . \tag{7.61}$$

Let M_0 be positive; the corresponding mass profile is the same as shown in Fig. 7.3a, with x replaced by y.

Now we look for solutions of Eq. (7.57) that reside in the energy range $-M_0 < E < M_0$, i.e., in the bulk gap of the two domains, and which propagate along, but decay perpendicular to, the domain wall at $y = 0$. Our wave-function Ansatz for the upper half plane $y > 0$ is

$$\varphi_u(x, y) = \begin{pmatrix} a_u \\ b_u \end{pmatrix} e^{iq_x x} e^{iq_y^{(u)} y} \tag{7.62}$$

with $q_x \in \mathbb{R}$, $q_y^{(u)} \in i\mathbb{R}^+$ and $a, b \in \mathbb{C}$. For the lower half plane, $\varphi_l(x, y)$ is defined as $\varphi_u(x, y)$ but with $u \leftrightarrow l$ interchanged and $q_y^{(l)} \in i\mathbb{R}^-$. The wave function $\varphi_u(x, y)$ does solve the two-dimensional Dirac equation defined by Eqs. (7.57) and (7.61) in the upper half plane $y > 0$ provided

$$q_y^{(u)} = i\kappa \equiv i\sqrt{M_0^2 + q_x^2 - E^2} \tag{7.63}$$

and

$$\frac{a_u}{b_u} = \frac{q_x + \kappa}{E - M_0} \tag{7.64}$$

Similar conditions apply for the ansatz $\varphi_l(x, y)$ for the lower half plane, with the substitutions $u \mapsto l$, $\kappa \mapsto -\kappa$ and $M_0 \mapsto -M_0$. The complete (unnormalized) envelope function has the form

$$\varphi(x, y) = \varphi_u(x, y)\Theta(y) + c\varphi_l(x, y)\Theta(-y), \tag{7.65}$$

where c is a yet unknown complex parameter to be determined from the boundary conditions at the domain wall.

The wave function (7.65) is an eigenstate of the EFA Hamiltonian with energy E if the boundary condition that the wave function is continuous on the line $y = 0$, that is,

$$\varphi_u(x, 0) = \varphi_l(x, 0), \tag{7.66}$$

is fulfilled for every x.

The boundary condition (7.66) determines the value of the parameter c as well as the dispersion relation $E(q_x)$ of the edge states. First, (7.66) implies

$$q_x - \kappa = c(q_x + \kappa) \Rightarrow c = \frac{q_x - \kappa}{q_x + \kappa}, \tag{7.67}$$

$$E + M_0 = c(E - M_0) \Rightarrow -q_x M_0 = \kappa E. \tag{7.68}$$

Note that κ depends on E according to Eq. (7.63). It is straightforward to find the dispersion relation of the edge states by solving $-q_x M_0 = \kappa(E)E$ for E with the condition $-M_0 < E < M_0$:

$$E = -q_x. \tag{7.69}$$

This simple dispersion relation is shown in Fig. 7.4a. Together with Eq. (7.63), this dispersion implies that the localization length of edge states is governed by M_0 only, i.e., is independent of q_x. The squared wave function of an edge state is shown in Fig. 7.4b.

A remarkable consequence of this simple dispersion relation is that the spinor components of the envelope function also have a simple form:

$$\begin{pmatrix} a_u \\ b_u \end{pmatrix} = \begin{pmatrix} a_l \\ b_l \end{pmatrix} = \begin{pmatrix} 1 \\ -1 \end{pmatrix}. \tag{7.70}$$

Edge states at similar mass domain walls at $u(y) \approx 0$ and $u(y) \approx 2$ can be derived analogously. Note that at $u(y) \approx 0$, the low-energy states can reside in two different Dirac valleys, around $\mathbf{k}_0 = (0, \pi)$ or $\mathbf{k}_0 = (\pi, 0)$, and there is one edge state in each valley. The number of edge states obtained in the continuum model, as well as their directions of propagation, are in correspondence with those obtained in the lattice model; as we have seen for the latter case, the number and direction are given by the magnitude and the sign of Chern-number difference across the domain wall, respectively.

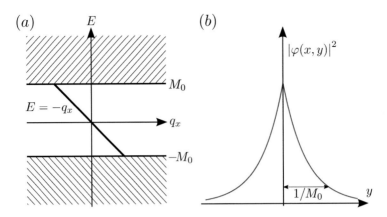

Fig. 7.4 Chiral state obtained from the two-dimensional Dirac equation. (**a**) Dispersion relation and (**b**) squared wave function of a chiral state confined to, and propagating along, a mass domain wall

An interesting fact is that the existence of the edge state is not constrained to case of a sharp, step-like domain wall described above. Moreover, the simple dispersion relation and spinor structure found above generalize for more irregular domain walls. This generalization is proven in a similar fashion as in the case of the SSH model, see Sect. 7.2.4. To see this, consider an almost arbitrary one-dimensional spatial dependence of the mass, similar to the one in Fig. 7.3b: $M(x, y) = M(y)$ with the only condition that M changes sign between the half-planes $y < 0$ and $y > 0$, i.e., $M(y \to -\infty) < 0$ and $M(y \to \infty) > 0$. We claim that there exists a solution of the corresponding two-dimensional Dirac equation that propagates along the domain wall, has the dispersion relation $E = -q_x$, is confined in the direction perpendicular to the domain wall, and has the wave function

$$\varphi(x, y) = \begin{pmatrix} 1 \\ -1 \end{pmatrix} e^{iq_x x} f(y). \tag{7.71}$$

To prove this proposition, insert this wave function $\varphi(x, y)$ to the two-dimensional Dirac equation and substitute E with $-q_x$ therein. This procedure results in two equivalent equations that are fulfilled if $\partial_y f(y) = -M(y) f(y)$, implying that Eq. (7.71) is indeed a normalizable solution with $E = -q_x$ provided that the function f has the form

$$f(y) = e^{-\int_0^y dy' M(y')}. \tag{7.72}$$

To summarize: In the preceding chapters, we introduced the topological characterization of lattice models and the corresponding edge states and states bound to domain walls between regions of different topological character. In this chapter, we demonstrated that a low-energy continuum description (the EFA Schrödinger equation) can be derived from a lattice model, and can be used to analyze those electronic states. Besides being a convenient analytical tool to describe inhomogeneous lattices, the envelope-function approximation also demonstrates that the emergence of topologically protected states is not restricted to lattice models.

Problems

7.1 SSH model with spatially dependent intracell hopping
In Sect. 7.2.3, we provide the EFA proposition for the SSH model with spatially dependent intracell hopping. Prove this proposition, following the procedure detailed in Sect. 7.1.3 for the one-dimensional monatomic chain.

7.2 SSH model with spatially dependent intercell hopping
Derive the EFA Hamiltonian for an inhomogeneous SSH model, where w varies in space and v is constant. Assume a nearly metallic scenario, $w(x) \approx v$.

7.3 QWZ model
Prove the EFA proposition for the QWZ model. The proposition is outlined in Sect. 7.3.3. The proof is analogous to that used for the SSH model.

7.4 QWZ model at $u \approx 2$
The bulk momentum-space Hamiltonian $\hat{H}(k)$ of the QWZ model is given in Eq. (6.1). Starting from this $\hat{H}(k)$, derive the EFA Hamiltonian describing low-energy excitations in the case of an inhomogeneous u parameter for which $u \approx 2$.

Chapter 8
Time-Reversal Symmetric Two-Dimensional Topological Insulators: The Bernevig–Hughes–Zhang Model

In the previous chapters, we have seen how two-dimensional insulators can host one-way propagating (a.k.a. chiral) edge states, which ensures reflectionless transport along the edge. The existence of chiral edge states precludes time-reversal symmetry. Indeed, time-reversed edge states would describe particles propagating backwards along the edge. In Chern insulators (two-dimensional insulators with nonvanishing Chern number), the absence of these counterpropagating states from the spectrum is what ensures the reflectionless propagation of particles along the edges.

What about time-reversal symmetric or time-reversal invariant two dimensional insulators? According to the above, they cannot be Chern insulators. Interestingly though, the same time-reversal symmetry that ensures that for every edge state mode there is a counterpropagating time reversed partner, can also ensure that no scattering between these two modes occurs. This means that it is possible for time-reversal invariant two-dimensional insulators to host edge states with reflectionless propagation, in both directions, at both edges. The details of why and how this happens are discussed in this and the following chapters.

We will find that all two-dimensional time-reversal invariant insulators fall into two classes: the trivial class, with an even number of pairs of edge states at a single edge, and the topological class, with an odd number of pairs of edge states at a single edge. We then subsequently show that disorder that breaks translational invariance along the edge can destroy edge state conduction in the trivial class, but not in the topological class.

The bulk–boundary correspondence for Chern insulators stated that the net number of edge states on the edge is the same as the Chern number of the bulk, Q. We showed this by mapping the two-dimensional system to a periodically, adiabatically pumped one-dimensional chain. After the mapping, the unit of charge pumped through the chain during a period could be identified with the net number of chiral edge states.

© Springer International Publishing Switzerland 2016
J.K. Asbóth et al., *A Short Course on Topological Insulators*, Lecture Notes in Physics 919, DOI 10.1007/978-3-319-25607-8_8

Unfortunately, identifying and calculating the bulk topological invariant of a time-reversal invariant two-dimensional insulator is much more cumbersome than for a Chern insulator. We therefore come back to this problem in the next chapter.

8.1 Time-Reversal Symmetry

Before we discuss time-reversal symmetric topological insulators, we first need to understand what we mean by time reversal symmetry, and how it leads to Kramers' degeneracy.

8.1.1 Time Reversal in Continuous Variable Quantum Mechanics (Without Spin)

Take a single particle with no internal degree of freedom, described by a wavefunction $\Psi(\mathbf{r})$. Its dynamics is prescribed by a time independent Hamiltonian $\hat{H} = (\hat{p} - e\mathbf{A}(\hat{r}))^2 + V(\hat{r})$, where the functions $\mathbf{A}$ and V are the vector and scalar potentials, respectively, and e is the charge of the particle. The corresponding Schrödinger equation for the wavefunction $\Psi(r, t)$ reads

$$i\partial_t \Psi(\mathbf{r}, t) = \left\{ (-i\partial_\mathbf{r} - e\mathbf{A}(\mathbf{r}))^2 + V(\mathbf{r}) \right\} \Psi(\mathbf{r}, t). \tag{8.1}$$

Any solution Ψ of the above equation can be complex conjugated, and gives a solution of the complex conjugate of the Schrödinger equation,

$$-i\partial_t \Psi(\mathbf{r}, t)^* = \left\{ (-i\partial_\mathbf{r} + e\mathbf{A}(\mathbf{r}))^2 + V(\mathbf{r}) \right\} \Psi(\mathbf{r}, t)^*. \tag{8.2}$$

8.1.1.1 The Operator of Complex Conjugation in Real Space Basis

We use K to denote the operator that complex conjugates everything to its right in real space basis,

$$Kf(\mathbf{r}) = f(\mathbf{r})^* K; \qquad\qquad K^2 = 1, \tag{8.3}$$

for any complex valued function $f(\mathbf{r})$ of position. The Schrödinger equation above can be rewritten using K as

$$Ki\partial_t \Psi = Ki\partial_t KK\Psi = -i\partial_t \Psi^* = K\hat{H}KK\Psi = \hat{H}^*\Psi^*. \tag{8.4}$$

Complex conjugation in real space basis conforms to intuitive expectations of time reversal: it is local in space, takes $\hat{x} \to \hat{x}$, and flips the momenta, $i\partial_x \to -i\partial_x$.

8.1.1.2 Time Reversal

The above relation shows that for any closed quantum mechanical system, there is a simple way to implement time reversal. This requires to change *both* the wavefunction Ψ to Ψ^* *and* the Hamiltonian $\hat{H}$ to $\hat{H}^*$. The change of the Hamiltonian involves $\mathbf{A} \to -\mathbf{A}$, i.e., flipping the sign of the vector potential.

8.1.1.3 Time Reversal Symmetry

In the special case where the Hamiltonian in real space basis is real, $\hat{H}^* = \hat{H}$, we can implement time reversal by only acting on the wavefunction. In that case, we say that the system has time reversal symmetry. For the scalar Schrödinger equation above, this happens if there is no vector potential, $\mathbf{A} = 0$. To see this more explicitly, consider time evolution for a time t, then apply the antiunitary operator K, then continue time evolution for time t, then apply K once more:

$$\hat{U} = Ke^{-i\hat{H}t}Ke^{-i\hat{H}t} = e^{-Ki\hat{H}tK}e^{-i\hat{H}t} = e^{i\hat{H}^*t}e^{-i\hat{H}t} \qquad (8.5)$$

If $\hat{H}^* = \hat{H}$, then $\hat{U} = 1$, which means that K acts like time reversal.

8.1.2 Lattice Models with an Internal Degree of Freedom

In these notes we deal with models for solids which are lattice Hamiltonians: the position (the external degree of freedom) is discrete, and there can be an internal degree of freedom (sublattice, orbital, spin, or other).

8.1.2.1 Definition of the Operator K of Complex Conjugation

For the operator of complex conjugation we need to fix not only the external position basis, $\mathscr{E}_{external} = \{|\mathbf{m}\rangle\}$, but also an internal basis, $\mathscr{E}_{internal} = \{|\alpha\rangle\}$. The property defining K then reads

$$\forall z \in \mathbb{C}, \quad \forall |\mathbf{m}\rangle, |\mathbf{m}'\rangle \in \mathscr{E}_{external}, \quad \forall |\alpha\rangle, |\alpha'\rangle \in \mathscr{E}_{internal}:$$

$$Kz|\mathbf{m}, \alpha\rangle\langle\mathbf{m}', \alpha'| = z^*|\mathbf{m}, \alpha\rangle\langle\mathbf{m}', \alpha'|K, \qquad (8.6)$$

where z^* is the complex conjugate of z.

We will use the shorthand $|\Psi^*\rangle$ and $\hat{A}^*$ to represent $K|\Psi\rangle$ and $K\hat{A}K$, respectively. The defining equations are

$$|\Psi\rangle = \sum_m \sum_\alpha \Psi_{m,\alpha} |m\rangle \otimes |\alpha\rangle ; \tag{8.7}$$

$$|\Psi^*\rangle = K|\Psi\rangle = \sum_m \sum_\alpha \Psi_{m,\alpha}^* |m\rangle \otimes |\alpha\rangle ; \tag{8.8}$$

$$\hat{A} = \sum_{m'm} \sum_{\alpha'\alpha} A_{m',\alpha',m,\alpha} |m'\rangle \langle m| \otimes |\alpha'\rangle \langle \alpha| ; \tag{8.9}$$

$$\hat{A}^* = K\hat{A}K = \sum_{m'm} \sum_{\alpha'\alpha} A_{m',\alpha',m,\alpha}^* |m'\rangle \langle m| \otimes |\alpha'\rangle \langle \alpha| . \tag{8.10}$$

8.1.2.2 Time-Reversal Affects External and Internal Degrees of Freedom

We look for a representation of time reversal symmetry $\hat{\mathcal{T}}$ in terms of a general antiunitary operator. Apart from complex conjugation, which acts on both external and internal Hilbert space, we allow for an additional unitary operation on the internal degrees of freedom $\hat{\tau}$, that is independent of position,

$$\hat{\mathcal{T}} = \hat{\tau} K. \tag{8.11}$$

We say that a Hamiltonian $\hat{H}$ is time reversal invariant (or time reversal symmetric) with respect to time reversal represented by $\hat{\mathcal{T}}$ if

$$\hat{\mathcal{T}} \hat{H} \hat{\mathcal{T}}^{-1} = \hat{H}. \tag{8.12}$$

In the same sense as for the chiral symmetry (cf. Sect. 1.4), when we talk about a Hamiltonian, what we really mean is a set of Hamiltonians $\hat{H}(\underline{\xi})$, with $\underline{\xi}$ representing parameters that are subject to disorder. Thus, Eq. (8.12) should hold for any of the $\hat{H}(\underline{\xi})$, with $\hat{\mathcal{T}}$ independent of $\underline{\xi}$.

8.1.3 Two Types of Time-Reversal

We can require that a time reversal operator $\hat{\mathcal{T}}$, when squared, should give at most a phase:

$$\hat{\tau} K \hat{\tau} K = \hat{\tau} \hat{\tau}^* = e^{i\phi} \mathbb{I}_{\text{internal}}. \tag{8.13}$$

If that was not the case, if the unitary operator $\hat{\tau}\hat{\tau}^*$ was nontrivial, then it would represent a unitary symmetry of a time-reversal invariant Hamiltonian, since

$$\hat{\tau}\hat{\tau}^*\hat{H}(\hat{\tau}\hat{\tau}^*)^\dagger = \hat{\tau}K\hat{\tau}K\hat{H}K\hat{\tau}^\dagger K\hat{\tau}^\dagger = \hat{\tau}K\hat{H}K\hat{\tau}^\dagger = \hat{H}. \tag{8.14}$$

As explained in Sect. 1.4, when we want to investigate topological phases, the usual first step is to get rid of unitary symmetries one by one (except for the lattice translation symmetry of the bulk Hamiltonian), by restricting our attention to a single superselection sector of each symmetry. Thus, the only time reversal symmetries that are left are those that fulfil Eq. (8.13).

The phase factor $e^{i\phi} = \hat{\mathcal{T}}^2$ turns out to have only two possible values: $+1$ or -1. Multiplying Eq. (8.13) from the left by $\hat{\tau}^\dagger$, we get $\hat{\tau}^* = e^{i\phi}\hat{\tau}^\dagger = e^{i\phi}(\hat{\tau}^*)^T$, where the superscript T denotes transposition. Iterating this last relation once more, we obtain $\hat{\tau}^* = e^{2i\phi}\hat{\tau}^*$, which means $e^{i\phi} = \pm 1$, wherefore

$$\hat{\mathcal{T}}^2 = \pm 1. \tag{8.15}$$

A Hamiltonian with no unitary symmetries can have only one type of time-reversal symmetry: either $\hat{\mathcal{T}}^2 = +1$, or $\hat{\mathcal{T}}^2 = -1$, but not both. Assume a Hamiltonian had two different time-reversal symmetries, $\hat{\mathcal{T}}$ and $\hat{\mathcal{T}}_1$. Along the lines of Eq. (8.14), the product of the two, the unitary operator $\hat{\mathcal{T}}_1\hat{\mathcal{T}}$ would then represent a unitary symmetry. The only exception is if $\hat{\mathcal{T}}_1 = e^{i\chi}\hat{\mathcal{T}}$, when they are not really different symmetries. However, in this case, since $\hat{\mathcal{T}}$ is antiunitary, these two symmetries square to the same number, $\hat{\mathcal{T}}_1^2 = e^{i\chi}\hat{\mathcal{T}}e^{i\chi}\hat{\mathcal{T}} = \hat{\mathcal{T}}^2$.

An example for a time-reversal operator with $\hat{\mathcal{T}}^2 = +1$ is given by the complex conjugation K. An example for a time-reversal operator with $\hat{\mathcal{T}}^2 = -1$ is time reversal for a spin-1/2 particle. Since time reversal should also flip the spin, it is achieved by $\hat{\mathcal{T}} = -i\sigma_y K$, with K defined on the basis of the eigenstates of $\hat{\sigma}_z$. The fact that this works can be checked by $\hat{\mathcal{T}}\hat{\sigma}_j\hat{\mathcal{T}}^{-1} = -\hat{\sigma}_j$ for $j = x, y, z$.

8.1.3.1 The Operator $\hat{\tau}$ Is Symmetric or Antisymmetric

Specifying the square of the time-reversal operation constrains the operator $\hat{\tau}$ to be symmetric or antisymmetric. Consider

$$\hat{\mathcal{T}}^2 = \hat{\tau}K\hat{\tau}K = \hat{\tau}\hat{\tau}^* = \pm 1; \qquad \hat{\tau}^* = \pm\hat{\tau}^\dagger = (\pm\hat{\tau}^T)^*, \tag{8.16}$$

where the subscript T denotes transpose in the same basis where the complex conjugate is defined. Therefore,

$$\hat{\mathcal{T}}^2 = +1 \quad \Longleftrightarrow \quad \hat{\tau} = \hat{\tau}^T \quad \text{symmetric;} \tag{8.17}$$

$$\hat{\mathcal{T}}^2 = -1 \quad \Longleftrightarrow \quad \hat{\tau} = -\hat{\tau}^T \quad \text{antisymmetric.} \tag{8.18}$$

8.1.4   Time Reversal of Type $\hat{\mathcal{T}}^2 = -1$ Gives Kramers' Degeneracy

A defining property of an antiunitary operator $\hat{\mathcal{T}}$ is that for any pair of states $|\Psi\rangle$ and $|\Phi\rangle$, we have

$$\langle \hat{\mathcal{T}}\Phi | \hat{\mathcal{T}}\Psi \rangle = (\hat{\tau}\,|\Phi^*\rangle)^\dagger \hat{\tau}\,|\Psi^*\rangle . = |\Phi^*\rangle^\dagger \hat{\tau}^\dagger \hat{\tau}\,|\Psi^*\rangle = \langle \Phi^* | \Psi^* \rangle = \langle \Phi | \Psi \rangle^* . \tag{8.19}$$

Consider now this relation with $|\Phi\rangle = \hat{\mathcal{T}}\,|\Psi\rangle$:

$$\langle \hat{\mathcal{T}}\Psi | \Psi \rangle^* = \langle \hat{\mathcal{T}}^2 \Psi | \hat{\mathcal{T}}\Psi \rangle = \langle \pm \Psi | \hat{\mathcal{T}}\Psi \rangle = \pm \langle \hat{\mathcal{T}}\Psi | \Psi \rangle^* , \tag{8.20}$$

where the $\pm$ stands for the square of the time reversal operator $\hat{\mathcal{T}}$, which is ± 1. If $\hat{\mathcal{T}}^2 = +1$, the above line gives no information, but if $\hat{\mathcal{T}}^2 = -1$, it leads immediately to $\langle \hat{\mathcal{T}}\Psi \,|\, \Psi \rangle = 0$, which means that for every energy eigenstate, its time-reversed partner, which is also an energy eigenstate with the same energy, is orthogonal. This is known as Kramers degeneracy.

8.1.5   Time-Reversal Symmetry of a Bulk Hamiltonian

We now calculate the effect of time-reversal symmetry $\hat{\mathcal{T}} = \hat{\tau}K$ on the bulk momentum-space Hamiltonian $\hat{H}(k)$. This latter is obtained, as in Sect. 1.2, by setting periodic boundary conditions, and defining a plane wave basis in the corresponding external Hilbert space as

$$|\mathbf{k}\rangle = \frac{1}{\sqrt{N_x N_y}} \sum_{\mathbf{k}} e^{i m \mathbf{k}}\,|\mathbf{m}\rangle ; \qquad \hat{\mathcal{T}}\,|\mathbf{k}\rangle = |-\mathbf{k}\rangle\,\hat{\mathcal{T}}. \tag{8.21}$$

Next, $\hat{H}_{\text{bulk}}$ is the part of $\hat{H}$ in the bulk, with periodic boundary conditions, whose components in the plane wave basis define the bulk momentum-space Hamiltonian,

$$\hat{H}(\mathbf{k}) = \langle \mathbf{k}|\,\hat{H}_{\text{bulk}}\,|\mathbf{k}\rangle ; \qquad \hat{H}_{\text{bulk}} = \sum_{\mathbf{k}} |\mathbf{k}\rangle\,\langle \mathbf{k}| \otimes \hat{H}(\mathbf{k}). \tag{8.22}$$

The effect of time-reversal symmetry follows,

$$\hat{\mathcal{T}}\hat{H}_{\text{bulk}}\hat{\mathcal{T}}^{-1} = \sum_{\mathbf{k}} |-\mathbf{k}\rangle\,\langle -\mathbf{k}| \otimes \hat{\tau}\hat{H}(\mathbf{k})^* \hat{\tau}^\dagger = \sum_{\mathbf{k}} |\mathbf{k}\rangle\,\langle \mathbf{k}| \otimes \hat{\tau}\hat{H}(-\mathbf{k})^* \hat{\tau}^\dagger. \tag{8.23}$$

We read off the action of $\hat{\mathcal{T}}$ on the bulk momentum-space Hamiltonian, and obtain the necessary requirement of time-reversal symmetry as

$$\hat{\tau}\hat{H}(-\mathbf{k})^*\hat{\tau}^\dagger = \hat{H}(\mathbf{k}). \tag{8.24}$$

Note that time-reversal symmetry of the bulk Hamiltonian is necessary, but not sufficient, for time-reversal symmetry of the system: perturbations at the edges can break time reversal.

A direct consequence of time-reversal symmetry is that the dispersion relation of a time-reversal symmetric Hamiltonian has to be symmetric with respect to inversion in the Brillouin zone, $\mathbf{k} \to -\mathbf{k}$. Indeed, take an eigenstate of $\hat{H}(\mathbf{k})$, with

$$\hat{H}(\mathbf{k})\,|u(\mathbf{k})\rangle = E(\mathbf{k})\,|u(\mathbf{k})\rangle . \tag{8.25}$$

Using time-reversal symmetry, Eq. (8.24), we obtain

$$\hat{\tau}\hat{H}(-\mathbf{k})^*\hat{\tau}^\dagger\,|u(\mathbf{k})\rangle = E(\mathbf{k})\,|u(\mathbf{k})\rangle . \tag{8.26}$$

Multiplying from the left by $\hat{\tau}^\dagger$ and complex conjugating, we have

$$\hat{H}(-\mathbf{k})\hat{\tau}^T\,|u(\mathbf{k})\rangle^* = E\hat{\tau}^T\,|u(\mathbf{k})\rangle^* . \tag{8.27}$$

This last line tells us that for every eigenstate $|u(\mathbf{k})\rangle$ of $\hat{H}(\mathbf{k})$, there is a time-reversed partner eigenstate of $\hat{H}(-\mathbf{k})$ at the same energy, $\hat{\tau}^T\,|u(\mathbf{k})\rangle^*$. This implies inversion symmetry of the energies, $E(\mathbf{k}) = E(-\mathbf{k})$. Note, however, that $E(\mathbf{k}) = E(-\mathbf{k})$ is not enough to guarantee time-reversal symmetry.

It is especially interesting to look at points in the Brillouin zone which map unto themselves under inversion: the Time-reversal invariant momenta (TRIM). In d dimensions there are 2^d such points, one of which is at the center of the Brillouin zone (so-called Γ point), and others at the boundary of the Brillouin zone. At such momenta, Eq. (8.24) implies that

$$\hat{\tau}\hat{H}(\mathbf{k}_{\text{TRIM}})^*\hat{\tau}^\dagger = \hat{H}(\mathbf{k}_{\text{TRIM}}). \tag{8.28}$$

If $\hat{\mathcal{T}}^2 = -1$, then because of Kramers degeneracy, at a time-reversal invariant momentum, every eigenvalue of the bulk momentum-space Hamiltonian is (at least) twice degenerate.

8.2 Doubling the Hilbert Space for Time-Reversal Symmetry

There is a simple way to construct a system with Time-Reversal Symmetry, $\hat{H}_{\text{TRI}}$, starting from a lattice Hamiltonian $\hat{H}$. We take two copies of the system, and change the Hamiltonian in one of them to $\hat{H}^* = K\hat{H}K$. We then couple them, much as we did to layer Chern insulators on top of each other in Sect. 6.2.4:

$$\hat{H}_{\text{TRI}} = |0\rangle \langle 0| \otimes \hat{H} + |1\rangle \langle 1| \otimes \hat{H}^* + \left(|0\rangle \langle 1| \otimes \mathbb{I}_{\text{external}} \otimes \hat{C} + h.c. \right), \quad (8.29)$$

where the hopping between the copies is accompanied by a position-independent operation $\hat{C}$ on the internal degree of freedom. In a matrix form, in real-space basis (and somewhat simplified notation), this reads

$$H_{\text{TRI}} = \begin{bmatrix} H & C \\ C^\dagger & H^* \end{bmatrix}. \quad (8.30)$$

We will use $\hat{s}_{x,y,z}$ to denote the Pauli operators acting on the "copy degree of freedom", defined as

$$\hat{s}_{x/y/z} = \hat{\sigma}_{x/y/z} \otimes \mathbb{I}_{\text{external}} \otimes \mathbb{I}_{\text{internal}}. \quad (8.31)$$

Using these operators, the time-reversal invariant Hamiltonian reads

$$\hat{H}_{\text{TRI}} = \frac{1 + \hat{s}_z}{2} \otimes \hat{H} + \frac{1 - \hat{s}_z}{2} \otimes \hat{H}^*$$
$$+ \frac{\hat{s}_x + i\hat{s}_y}{2} \otimes \mathbb{I}_{\text{external}} \otimes \hat{C} + \frac{\hat{s}_x - i\hat{s}_y}{2} \otimes \mathbb{I}_{\text{external}} \otimes \hat{C}^\dagger. \quad (8.32)$$

The choice of the coupling operator $\hat{C}$ is important, as it decides which type of time-reversal symmetry $\hat{H}_{\text{TRI}}$ will have.

8.2.1 Time Reversal with $\hat{\mathcal{T}}^2 = -1$ Requires Antisymmetric Coupling Operator $\hat{C}$

If we want a time-reversal symmetry that squares to -1, we can go for

$$\hat{\mathcal{T}} = i\hat{s}_y K; \qquad\qquad \hat{\mathcal{T}}^{-1} = K(-i)\hat{s}_y, \quad (8.33)$$

with the factor of i is included for convenience, so that the matrix of $i\hat{s}_y$ is real. The requirement of time-reversal symmetry can be obtained using

$$(i\hat{s}_y K)H_{\text{TRI}}(i\hat{s}_y K)^{-1} = \begin{bmatrix} 0 & 1 \\ -1 & 0 \end{bmatrix} \cdot \begin{bmatrix} H^* & C^* \\ C^T & H \end{bmatrix} \cdot \begin{bmatrix} 0 & -1 \\ 1 & 0 \end{bmatrix}$$
$$= \begin{bmatrix} C^T & H \\ -H^* & -C^* \end{bmatrix} \cdot \begin{bmatrix} 0 & -1 \\ 1 & 0 \end{bmatrix} = \begin{bmatrix} H & -C^T \\ -C^* & H^* \end{bmatrix}. \quad (8.34)$$

We have time-reversal symmetry represented by $\hat{\mathscr{T}} = i\hat{s}_y K$, if

$$i\hat{s}_y K \hat{H}_{\text{TRI}} (i\hat{s}_y K)^{-1} = \hat{H}_{\text{TRI}} \qquad \Leftrightarrow \qquad \hat{C} = -\hat{C}^T, \tag{8.35}$$

where the subscript T denotes transposition in the same fixed internal basis that is used to define complex conjugation K.

8.2.2 Symmetric Coupling Operator $\hat{C}$ Gives Time Reversal with $\hat{\mathscr{T}}^2 = +1$

If the coupling operator is symmetric, $\hat{C} = \hat{C}^T$, then the same derivation as above shows that we have time-reversal symmetry represented by $\hat{\mathscr{T}} = \hat{s}_x K$,

$$\hat{s}_x K \hat{H}_{\text{TRI}} K \hat{s}_x = \hat{H}_{\text{TRI}} \qquad \Leftrightarrow \qquad \hat{C} = \hat{C}^T. \tag{8.36}$$

This time-reversal operator squares to $+1$.

If all we want is a lattice Hamiltonian with a time-reversal symmetry that squares to $+1$, we don't even need to double the Hilbert space. We can just take

$$\hat{\mathscr{T}} = K; \qquad \hat{H}_{\text{TRI}} = \frac{\hat{H} + \hat{H}^*}{2} = K\hat{H}_{\text{TRI}} K. \tag{8.37}$$

Colloquially, this construction is referred to as taking the real part of the Hamiltonian.

8.3 A Concrete Example: The Bernevig-Hughes-Zhang Model

To have an example, we use the construction above to build a toy model—called Bernevig-Hughes-Zhang (BHZ) model—for a time-reversal invariant topological insulator starting from the QWZ model of Chap. 6. We follow the construction through using the bulk momentum-space Hamiltonian, and obtain

$$\hat{H}_{\text{BHZ}}(\mathbf{k}) = \hat{s}_0 \otimes [(u + \cos k_x + \cos k_y)\hat{\sigma}_z + \sin k_y \hat{\sigma}_y)] + \hat{s}_z \otimes \sin k_x \hat{\sigma}_x + \hat{s}_x \otimes \hat{C}, \tag{8.38}$$

where $\hat{C}$ is a Hermitian coupling operator acting on the internal degree of freedom. For $\hat{C} = 0$, the Hamiltonian $\hat{H}_{\text{BHZ}}$ reduces to the 4-band toy model for HgTe, introduced by Bernevig, Hughes and Zhang [5].

8.3.1 Two Time-Reversal Symmetries If There Is No Coupling

If there is no coupling between the copies, $\hat{C} = 0$, the BHZ model has two different time-reversal symmetries, $\hat{\mathcal{T}} = i\hat{s}_y K$ and $\hat{\mathcal{T}}_2 = \hat{s}_x K$, due to its block diagonal structure reflecting a unitary symmetry, $\hat{s}_z \hat{H}_{\mathrm{BHZ}} \hat{s}_z^\dagger = \hat{H}_{\mathrm{BHZ}}$. In this situation, the type of predictions we can make will depend on which of these symmetries is robust against disorder. We will in the following require the $\hat{\mathcal{T}}^2 = -1$ symmetry. If this symmetry is robust, then everything we do will apply to the $\hat{C} = 0$ case of $\hat{H}_{\mathrm{BHZ}}$. The extra time reversal symmetry in that case is just a reminder that most features could be calculated in a more simple way, by working in the superselection sectors of $\hat{s}_z$ separately.

8.4 Edge States in Two-Dimensional Time-Reversal Invariant Insulators

We now consider the situation of edge states in a two-dimensional lattice Hamiltonian with time-reversal symmetry, much in the same way as we did for Chern insulators in Sect. 6.2.

8.4.1 An Example: The BHZ Model with Different Types of Coupling

We start with the concrete example of the BHZ model. We set the sublattice potential parameter $u = -1.2$, and plot the edge dispersion relation, defined in the same way as for the Chern insulators in Sect. 6.2.

As long as there is no coupling between the two copies, $\hat{C} = 0$, the system $\hat{H}_{\mathrm{BHZ}}$ is a direct sum of two Chern insulators, with opposite Chern numbers. As Fig. 8.1a shows, on each edge, there is a pair of edge state branches: a branch on the layer with Hamiltonian $\hat{H}$, and a counterpropagating branch on the layer with $\hat{H}^*$. Although these two edge state branches cross, this crossing will turn into an anticrossing: the states cannot scatter into each other since they are on different layers. The two edge state branches are linked by time-reversal: they occupy the same position, but describe propagation in opposite directions. In fact, they are linked by both time-reversal symmetries this system has, by $\hat{s}_x K$ and $i\hat{s}_y K$.

A coupling between the layers can gap the edge states out, as shown in Fig. 8.1b. We here used $\hat{C} = 0.3\hat{\sigma}_x$, which respects the $\hat{\mathcal{T}}^2 = +1$ symmetry but breaks the $\hat{\mathcal{T}}^2 = -1$ one. The crossings between counterpropagating edge states on the same edge have turned into anticrossings, as expected, since the coupling allows particles to hop between the counterpropagating edge states (on the same edge, but in different layers).

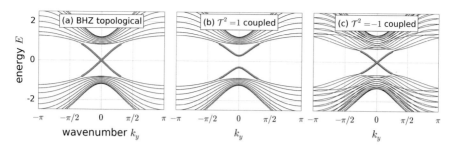

Fig. 8.1 Stripe dispersion relations of the BHZ model with stripe width $N = 10$, and with sublattice potential parameter $u = -1.2$. Right/left edge states (more than 60 % weight on the last/first two columns of unit cells) marked in dark red/light blue. (**a**): uncoupled layers, $\hat{C} = 0$. (**b**): Symmetric coupling $\hat{C} = 0.3\hat{\sigma}_x$ gaps the edge states out. (**c**): Antisymmetric coupling $\hat{C} = 0.3\hat{\sigma}_y$ cannot open a gap in the edge spectrum

We see something different if we couple the layers while respecting the $\hat{\mathscr{T}}^2 = -1$ time reversal symmetry, by, e.g., $\hat{C} = 0.3\sigma_y$. As Fig. 8.1c shows, the crossing at $k_y = 0$ between the edge state branches now does not turn into an anticrossing. As long as the coupling is not strong enough to close the bulk gap, the edge states here appear to be *protected*.

8.4.2 Edge States in $\hat{\mathscr{T}}^2 = -1$

The states form one-dimensional edge state bands in the one-dimensional Brillouin zone $k_x = -\pi, \ldots, \pi$, shown schematically in Fig. 8.2. In general, an edge will host edge states propagating in both directions. However, due to time-reversal symmetry, the dispersion relations must be left-right symmetric when plotted against the wavenumber k_x along the edge direction. This means that the number N_+ of right-moving edge states (these are plotted with solid lines in Fig. 8.2), and N_-, the number of left-moving edge states (dashed lines) have to be equal at any energy,

$$N_+(E) = N_-(E). \tag{8.39}$$

As with Chern insulators, we next consider the effect of adiabatic deformations of the clean Hamiltonian on edge states. We consider terms in the Hamiltonian that conserve translational invariance along the edge, and respect Time Reversal Symmetry. The whole discussion of Sect. 6.3 applies, and therefore adiabatic deformations cannot change the signed sum of the left- and right-propagating edge states in the gap. However, time-reversal symmetry restricts this sum to zero anyway.

Time reversal symmetry that squares to $\hat{\mathscr{T}}^2 = -1$, however, provides a further restriction: adiabatic deformations can only change the number of edge states by integer multiples of four (pairs of pairs). To understand why, consider the adiabatic

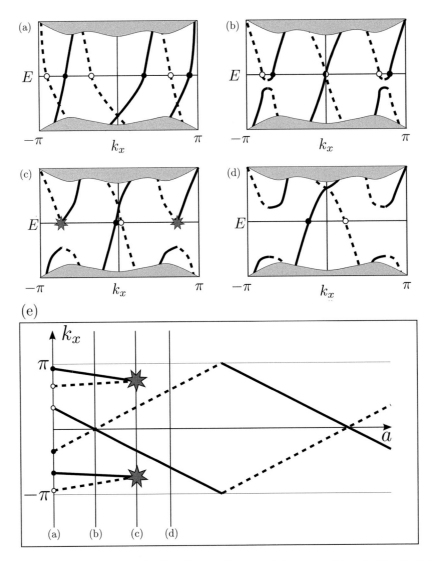

Fig. 8.2 Edge states on a single edge of a two-dimensional time-reversal invariant topological insulator with $\hat{\mathcal{T}}^2 = -1$, as the edge region undergoes a continuous deformation, parametrized by a, respecting the symmetry and the translational invariance along the edge. In (**a**)–(**d**), the edge state dispersions relations are shown, in the full edge Brillouin zone $k_x = -\pi, \ldots, \pi$, and in the energy window corresponding to the bulk gap. For clarity, right- (left-) propagating edge states are denoted by continuous (dashed) lines. Due to the deformation of the Hamiltonian, the edge state branches can move, bend, and couple, while the bulk remains unchanged. From (**a**) to (**b**), the crossing points between counterpropagating edge states become anticrossings, i.e., gaps open in these pairs of dispersion relation branches as a usual consequence of any parameter coupling them. Crossings at $k_x = 0$ and $k_x = \pi$ cannot be gapped, as this would lead to a violation of the Kramers theorem. From (**b**) to (**d**), these gaps become so large that at energy $E = 0$, the number of edge states drops from 6 (3 Kramers pairs) to 2 (1 Kramers pair). The k_x values of the edge states at zero energy are plotted in (**e**), where this change in the number of edge states shows up as an "annihilation" of right-propagating and left-propagating edge states

deformation corresponding to Fig. 8.2a–d. Degeneracies in the dispersion relation can be lifted by coupling the edge states, as it happens in (b), and this can lead to certain edge states disappearing at certain energies, as in (c). This can be visualized by plotting the k_x values at $E = 0$ of the branches of the edge state dispersion as functions of the deformation parameter (which is some combination of the parameters of the Hamiltonian) a, as in Fig. 8.2e. Due to the deformation, two counterpropagating edge states can "annihilate", when the corresponding modes form an avoided crossing. If this happens at a generic momentum value k, as in (c), then, due to the time reversal invariance, it also has to happen at $-k$, and so the number of edge states decreases by 4, not by 2. The special momentum values of k_x are the time-reversal invariant momenta, which in this case are $k_x = 0, \pm\pi$. If the edge state momenta meet at a time-reversal invariant momentum, as in (b) at $k_x = 0$, their "annihilation" would change the number of edge states by 2 and not by 4. However, this cannot happen, as it would create a situation that violates the Kramers degeneracy: at the time-reversal invariant momenta, energy eigenstates have to be doubly degenerate. The deformations in Fig. 8.2 can be also read from (d) to (a), and therefore apply to the introduction of new edge states as well.

8.4.3 $\mathbb{Z}_2$ Invariant: Parity of Edge State Pairs

At any energy inside the bulk gap, the parity of the number of edge-state Kramers pairs for a given dispersion relation is well defined. In Fig. 8.2a, there are 3 edge-state Kramers pairs for any energy in the bulk gap, i.e., the parity is odd. In Fig. 8.2c, there are 3 of them for every energy except for energies in the mini-gap of the bands on the left and right for which the number of edge-state Kramers pairs is 1, and for the upper and lower boundaries of the mini-gap [the former depicted by the horizontal line in Fig. 8.2c], where the number of Kramers pairs is 2. The parity is odd at almost every energy, except the two isolated energy values at the mini-gap boundaries.

The general proposition is that the parity of the number of edge-state Kramers pairs at a given edge for a given Hamiltonian at a given energy is independent of the choice of energy, as long as this energy is inside the bulk gap. Since in a time-reversal invariant system, all edge states have counter-propagating partners, we can express this number as

$$D = \frac{N(E)}{2} \mod 2 = \frac{N_+(E) + N_-(E)}{2} \mod 2, \tag{8.40}$$

where $N(E) = N_+(E) + N_-(E)$ is the total number of edge states at an edge. A caveat is that there are a few isolated energy values where this quantity is not well defined, e.g., the boundaries of mini-gaps in the above example, but these energies form a set of zero measure.

Since D is a topological invariant, we can classify two-dimensional time-reversal invariant lattice models according to it, i.e., the parity of the number of edge-state

Kramers pairs supported by a single edge of the terminated lattice. Because it can take on two values, this 'label' D is called the $\mathbb{Z}_2$ invariant, and is represented by a bit taking on the value 1 (0) if the parity is odd (even).

As a final step, we should next consider disorder that breaks translational invariance along the edge, in the same way as we did for Chern insulators. Due to the presence of edge states propagating in both directions along the edge, the treatment of disorder is a bit trickier than it was for chiral edge states.

8.5 Absence of Backscattering

A remarkable property of Chern insulators is that they support chiral edge states, i.e., edge states that have no counter-propagating counterparts. A simple fact implied by the chiral nature of these edge states is that impurities are unable to backscatter particles. As we argue below, absence of backscattering is also characteristic of disordered two-dimensional time-reversal invariant topological insulators with $\hat{\mathcal{T}}^2 = -1$, although the robustness is guaranteed only against time-reversal symmetric scatterers.

Here, we introduce the scattering matrix, a concept that allows for a formal analysis of scattering at impurities, and discuss the properties of edge state scattering in two-dimensional time-reversal invariant topological insulators with $\hat{\mathcal{T}}^2 = -1$. The scattering matrix will also serve as a basic tool in the last chapter, where we give a theoretical description of electronic transport of phase-coherent electrons, and discuss observable consequences of the existence and robustness of edge states.

8.5.1 The Scattering Matrix

Consider a phase-coherent two-dimensional conductor with a finite width in the y direction, and discrete translational invariance along the x axis. Think of the system as having periodic boundary conditions in the x direction. As earlier, we describe the system in terms of a simple lattice model, where the unit cells form a square lattice of size $N_x \times N_y$, and there might be an internal degree of freedom associated to the unit cells.

As the system has discrete translational invariance along x, we can also think of it as a one-dimensional lattice, whose unit cell incorporates both the internal degree of freedom of the two-dimensional lattice and the real-space structure along the y axis. Using that picture, it is clear that the electronic energy eigenstates propagating along the x axis at energy E have a product structure, as required by the one-dimensional Bloch's theorem:

$$|l, \pm\rangle = |k_{l,\pm}\rangle \otimes |\Phi_{l,\pm}\rangle , \qquad (8.41)$$

On the right hand side, the first ket corresponds to a usual momentum eigenstate $|k\rangle = \frac{1}{\sqrt{N_x}} \sum_{m_x=1}^{N_x} e^{ikm_x} |m_x\rangle$ propagating along x, whereas the second ket incorporates the shape of the transverse standing mode as well as the internal degree of freedom. The states appearing in Eq. (8.41) are normalized to unity. The integer $l = 1, 2, \ldots, N$ labels the propagating modes, also referred to as scattering channels. The $+$ and $-$ signs correspond to right-moving and left-moving states, respectively; the direction of movement is assigned according to the sign of the group velocity:

$$v_{l,\pm} = \left. \frac{dE(k)}{dk} \right|_{k=k_{l,\pm}}, \qquad (8.42)$$

where $E(k)$ is the dispersion relation of the one-dimensional band hosting the state $|l, \pm\rangle$.

Now we re-normalize the states above, such that different states carry the same particle current through an arbitrary vertical cross section of the system. We will use these current-normalized wave functions in the definition of the scattering matrix below, which guarantees that the latter is a unitary matrix. According to the one-dimensional relation we obtained between the current and the group velocity in Eq. (5.13), the current-normalized states can be defined using the group velocity as

$$|l, \pm\rangle_c = \frac{1}{\sqrt{|v_{l,\pm}|}} |l, \pm\rangle . \qquad (8.43)$$

Now consider the situation when the electrons are obstructed by a disordered region in the conductor, as shown in Fig. 8.3. A monoenergetic wave incident on the scattering region is characterized by a vector of coefficients

$$a^{(\text{in})} = \left(a_{L,1}^{(\text{in})}, a_{L,2}^{(\text{in})}, \ldots, a_{L,N}^{(\text{in})}, a_{R,1}^{(\text{in})}, a_{R,2}^{(\text{in})}, \ldots, a_{R,N}^{(\text{in})} \right). \qquad (8.44)$$

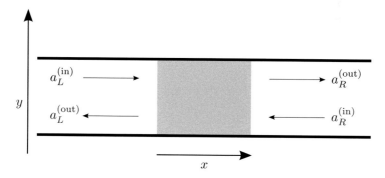

Fig. 8.3 Disordered region (gray) obstructing electrons in a two-dimensional phase-coherent conductor. The scattering matrix S relates the amplitudes $a_L^{(\text{in})}$ and $a_R^{(\text{in})}$ of incoming waves to the amplitudes $a_L^{(\text{out})}$ and $a_R^{(\text{out})}$ of outgoing waves

The first (second) set of N coefficients correspond to propagating waves (8.43) in the left (right) *lead L (R)*, that is, the clean regions on the left (right) side of the disordered region. The reflected and transmitted parts of the wave are described by the vector

$$a^{(\text{out})} = \left(a_{L,1}^{(\text{out})}, a_{L,2}^{(\text{out})}, \ldots, a_{L,N}^{(\text{out})}, a_{R,1}^{(\text{out})}, a_{R,2}^{(\text{out})}, \ldots, a_{R,N}^{(\text{out})} \right). \tag{8.45}$$

The corresponding energy eigenstate reads

$$|\psi\rangle = \sum_{l=1}^{N} a_{L,l}^{(\text{in})} |l, +, L\rangle_c + a_{L,l}^{(\text{out})} |l, -, L\rangle_c + a_{R,l}^{(\text{in})} |l, -, R\rangle_c + a_{R,l}^{(\text{out})} |l, +, R\rangle_c.$$
$$\tag{8.46}$$

Here, the notation for the current-normalized states introduced in Eq. (8.43) has been expanded by the lead index L/R.

The scattering matrix S relates the two vectors introduced in Eqs. (8.44) and (8.45):

$$a^{(\text{out})} = S a^{(\text{in})}. \tag{8.47}$$

The size of the scattering matrix is $2N \times 2N$, and it has the following block structure:

$$S = \begin{pmatrix} r & t' \\ t & r' \end{pmatrix} \tag{8.48}$$

where r and r' are $N \times N$ reflection matrices describing reflection from left to left and from right to right, and t and t' are transmission matrices describing transmission from left to right and right to left.

Particle conservation, together with the current normalization Eq. (8.43), implies the unitarity of the scattering matrix S. In turn, its unitary character implies that the Hermitian matrices $tt^\dagger$, $t't'^\dagger$, $1 - rr^\dagger$, and $1 - r'r'^\dagger$ all have the same set of real eigenvalues $T_1, T_2, \ldots T_N$, called transmission eigenvalues.

8.5.2   A Single Kramers Pair of Edge States

Now we use the scattering matrix S to characterize defect-induced scattering of an electron occupying an edge state of a two-dimensional time-reversal invariant topological insulator. Consider a half-plane of such a homogeneous lattice which supports exactly one Kramers pair of edge states at a given energy E in the bulk gap, as shown in Fig. 8.4. Consider the scattering of the electron incident on the defect from the left side in Fig. 8.4. The scatterer is characterized by the Hamiltonian V. We will show that the impurity cannot backscatter the electron as long as V is time-reversal symmetric.

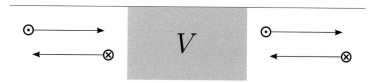

Fig. 8.4 Scattering of an edge state on a time-reversal symmetric defect V. In a two-dimensional time-reversal invariant topological insulator with $\hat{\mathscr{T}}^2 = -1$, having a single Kramers pair of edge states, the incoming electron is transmitted through such a defect region with unit probability

We start out without specifying the number of propagating edge-state Kramers pairs on the edge. Later we distinguish between the cases when (i) that number is one, and (ii) when that is a higher odd number. We choose our propagating modes such that the incoming and outgoing states are related by time-reversal symmetry, i.e.,

$$|l, -, L\rangle_c = \hat{\mathscr{T}} |l, +, L\rangle_c \tag{8.49a}$$

$$|l, +, R\rangle_c = \hat{\mathscr{T}} |l, -, R\rangle_c . \tag{8.49b}$$

Also, recall that $\hat{\mathscr{T}}^2 = -1$. In the presence of the perturbation V, the edge states of the disorder-free system are no longer energy eigenstates of the system. A general scattering state $|\psi\rangle$ at energy E is characterized by the vector $a^{(\text{in})}$ of incoming amplitudes. According to Eq. (8.46) and the definition (8.47) of the scattering matrix S, the energy eigenstates outside the scattering region can be expressed as:

$$|\psi\rangle = \sum_{l=1}^{N} \Big[a_{L,l}^{(\text{in})} |l, +, L\rangle_c + a_{R,l}^{(\text{in})} |l, -, R\rangle_c \tag{8.50}$$

$$+ \left(S a^{(\text{in})} \right)_{L,l} |l, -, L\rangle_c + \left(S a^{(\text{in})} \right)_{R,l} |l, +, R\rangle_c \Big].$$

Using Eq. (8.49), we find

$$-\hat{\mathscr{T}} |\psi\rangle = \sum_{l=1}^{N} \Big[-a_{L,l}^{(\text{in})*} |l, -, L\rangle_c - a_{R,l}^{(\text{in})*} |l, +, R\rangle_c \tag{8.51}$$

$$+ \left(S^* a^{(\text{in})*} \right)_{L,l} |l, +, L\rangle_c + \left(S^* a^{(\text{in})*} \right)_{R,l} |l, -, R\rangle_c \Big].$$

Due to time reversal symmetry, this state $-\hat{\mathscr{T}} |\psi\rangle$ is also an energy eigenstate having the same energy as $|\psi\rangle$. Using the unitary character of the scattering matrix, the state $-\hat{\mathscr{T}} |\psi\rangle$ can be rewritten as

$$-\hat{\mathscr{T}} |\psi\rangle = \sum_{l=1}^{N} \Big[\left(S^* a^{(\text{in})*} \right)_{L,l} |l, +, L\rangle_c + \left(S^* a^{(\text{in})*} \right)_{R,l} |l, -, R\rangle_c \tag{8.52}$$

$$+ \left(-S^T S^* a^{(\text{in})*} \right)_{L,l} |l, -, L\rangle_c + \left(-S^T S^* a^{(\text{in})*} \right)_{R,l} |l, +, R\rangle_c \Big].$$

Here, S^T denotes the transpose of S, is also an energy eigenstate having the same energy as $|\psi\rangle$. Comparing Eqs. (8.50) and (8.52), and knowing that the scattering matrix at a given energy is uniquely defined, we conclude that $S = -S^T$, that is

$$\begin{pmatrix} r & t' \\ t & r' \end{pmatrix} = S = -S^T = \begin{pmatrix} -r & -t \\ -t' & -r' \end{pmatrix}. \tag{8.53}$$

If we have a single edge-state Kramers pair, then the latter relation implies

$$r = r' = 0, \tag{8.54}$$

and hence perfect transmission of each of the two incoming waves.

   If the lattice has the geometry of a ribbon, and the time-reversal symmetric scatterer extends to both edges, then the absence of backscattering is not guaranteed. This is illustrated in Fig. 8.5, where we compare three examples. In (a), the defect is formed as a wide constriction on both edges, with a width much larger than the characteristic length of the penetration of the edge states to the bulk region of the ribbon. Backscattering between states at the same edge is forbidden due to time-reversal symmetry, and backscattering between states at different edges is forbidden due to a large spatial separation of their corresponding wave functions. In (b), a similar but narrower constriction with a width comparable to the penetration length of the edge states does allow for scattering between states on the lower and upper edges. In this case, backscattering from a right-moving state on one edge to a left-moving state at the other edge is not forbidden. In (c), the constriction divides

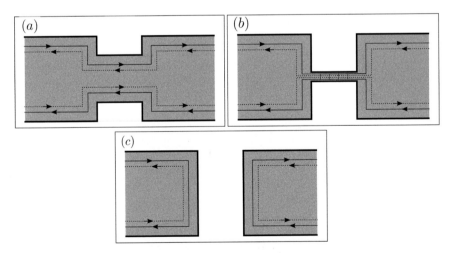

Fig. 8.5 Backscattering of edge states at a constriction. The states forming the edge-state Kramers pairs are depicted as solid and dashed lines. (**a**) A time-reversal symmetric defect localized to the edges, such as a small constriction shown here, is unable to backscatter the incoming electron. (**b**) Backscattering is possible between different edges, if the width of the constriction is of the order of the decay length of the edge states. (**c**) A finite spatial gap between the left and right part of the wire implies zero transmission

the ribbon to two unconnected parts, resulting in zero transmission through the constriction.

Naturally, backscattering is also allowed if the scatterer is not time-reversal symmetric, or if the scattering process is inelastic. Backscattering is not forbidden for the 'unprotected' edge states of topologically trivial ($D = 0$) two-dimensional time-reversal invariant insulators.

8.5.3 An Odd Number of Kramers Pairs of Edge States

The above statement (8.54) implying unit transmission can be generalized for arbitrary two-dimensional time-reversal invariant topological insulator lattice models, including those where the number of edge-state Kramers pairs N is odd but not one. The proposition is that in such a system, given a time-reversal symmetric scatterer V and an arbitrary energy E in the bulk gap, there exists at least one linear combination of the incoming states of energy E from each side of the defect that is perfectly transmitted through the defect.

The proof follows that in the preceding section, with the difference that the quantities r and t describing reflection and transmission are $N \times N$ matrices, and that the antisymmetric nature of the S-matrix $S = -S^T$ implies the antisymmetry of the reflection matrices $r = -r^T$. According to Jacobi's theorem, every odd-dimensional antisymmetric matrix has a vanishing determinant, which is implied by

$$\det(r) = \det(r^T) = \det(-r) = (-1)^N \det(r) = -\det(r), \tag{8.55}$$

where we used the antisymmetry of r in the second step and the oddness of N in the last step. As a consequence of Eq. (8.55) we know $\det(r) = 0$, hence $\det(r^\dagger r) = \det(r^\dagger)\det(r) = 0$. Therefore, at least one eigenvalue of $r^\dagger r$ is zero, which implies that at least one transmission eigenvalue T_l is unity.

8.5.4 Robustness Against Disorder

The absence-of-backscattering result (8.54) implies a remarkable statement regarding the existence of (at least) one perfectly transmitting edge state in a finite-size disordered sample of a two-dimensional time-reversal invariant topological insulator. (See also the discussion about Fig. 6.9 in the context of Chern insulators.) Such a sample with an arbitrarily chosen geometry is shown in Fig. 8.6. Assume that the disorder is time-reversal symmetric and localized to the edge of the sample. We claim that any chosen segment of the edge of this disordered sample supports, at any energy that is deep inside the bulk gap, (at least) one counterpropagating Kramers pair of edge states that are delocalized along the edge and able to transmit electrons with unit probability. This is a rather surprising feature in light of the fact that in truly

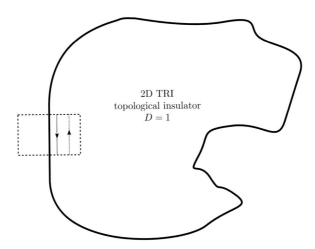

Fig. 8.6 A disordered two-dimensional time-reversal invariant topological insulator contacted with two electrodes. Disorder is 'switched off' and the edge is 'straightened out' within the dashed box, hence the edge modes there resemble those of the disorder-free lattice

one-dimensional lattices, a small disorder is enough to induce Anderson localization of the energy eigenstates, and hence render the system an electronic insulator.

To demonstrate the above statement, let us choose an edge segment of the disordered sample for consideration, e.g., the edge segment running outside the dashed box in Fig. 8.6. Now imagine that we 'switch off' disorder in the complementer part of the edge of the sample, and 'straighten out' the geometry of that complementer part, the latter being shown within the dashed box of Fig. 8.6. Furthermore, via an appropriate spatial adiabatic deformation of the Hamiltonian of the system in the vicinity of the complementer part of the edge (i.e., within the dashed box in Fig. 8.6), we make sure that only a single edge-state Kramers pair is present within this complementer part. The existence of such an adiabatic deformation is guaranteed by the topologically nontrivial character of the sample, see the discussion of Fig. 8.2. The disordered edge segment, outside the dashed box in Fig. 8.6, now functions as a scattering region for the electrons in the straightened part of the edge. From the result (8.54) we know that such a time-reversal symmetric scatterer is unable to induce backscattering between the edge modes of the straightened part of the edge, hence we must conclude that the disordered segment must indeed support a perfectly transmitting edge state in each of the two propagation directions.

In the last chapter, we show that the electrical conductance of such a disordered sample is finite and 'quantized', if it is measured through a source and a drain contact that couple effectively to the edge states.

Chapter 9
The $\mathbb{Z}_2$ Invariant of Two-Dimensional Topological Insulators

In the previous chapter, we have seen that two-dimensional insulators can host topologically protected edge states even if time-reversal symmetry is not broken, provided it squares to -1, i.e., $\hat{\mathcal{T}}^2 = -1$. Such systems fall into two categories: no topologically protected edge states (trivial), or one pair of such edge states (topological). This property defines a $\mathbb{Z}_2$ invariant for these insulators. In the spirit of the bulk–boundary correspondence, we expect that the bulk momentum-space Hamiltonian $\hat{H}(\mathbf{k})$ should have a corresponding topological invariant (generalized winding number).

The bulk $\mathbb{Z}_2$ invariant is notoriously difficult to calculate. The original definition of the invariant [11, 19] uses a smooth gauge in the whole Brillouin zone, that is hard to construct [31], which makes the invariant difficult to calculate. An altogether different approach, which is robust and calculatable, uses the scattering matrix instead of the Hamiltonian [13].

In this chapter we review a definition of the bulk $\mathbb{Z}_2$ invariant [37] based on the dimensional reduction to charge pumps. This is equivalent to the originally defined bulk invariants [37], but no smooth gauge is required to calculate it. It can be outlined as follows.

1. Start with a bulk Hamiltonian $\hat{H}(k_x, k_y)$. Reinterpret k_y as time: $\hat{H}(k_x, k_y)$ can be thought of as a bulk one-dimensional Hamiltonian of an adiabatic pump. This is the dimensional reduction we used for Chern insulators in Chap. 6.
2. Track the motion, with k_y playing the role of time, of pumped particles in the bulk using Wannier states. We will call this the Wannier center flow.
3. Time-reversal symmetry restricts the Wannier center flow. As a result, in some cases the particle pump cannot be turned off adiabatically—in those cases the insulator is topological. If it can be turned off, the insulator is trivial.

In order to go through this argument, we will first gather the used mathematical tools, i.e., generalize the Berry phase and the Wannier states. We will then define the Wannier center flow, show that it can be calculated from the Wilson loop. Finally,

© Springer International Publishing Switzerland 2016
J.K. Asbóth et al., *A Short Course on Topological Insulators*, Lecture Notes
in Physics 919, DOI 10.1007/978-3-319-25607-8_9

we will use examples to illustrate the $\mathbb{Z}_2$ invariant and argue that it gives the number of topologically protected edge states.

9.1 Tools: Nonabelian Berry Phase, Multiband Wannier States

To proceed to calculate the topological invariants, we need to generalize the tools of geometric phases, introduced in Chap. 2, and of the Wannier states of Chap. 3, to manifolds consisting of more bands.

9.1.1 Preparation: Nonabelian Berry Phase

We defined the Berry phase in Chap. 2, as the relative phase around a loop L of N states $|\Psi_j\rangle$, with $j = 1, 2, \ldots, N$. Since the Berry phase is gauge independent, it is really a property of the loop over N one-dimensional projectors $|\Psi_j\rangle\langle\Psi_j|$. In most physical applications—in our case as well—the elements of the loop are specified as projectors to a eigenstates of some Hamiltonian for N different settings of some parameters.

As a generalization of the Berry phase, we ask about the relative phase around a loop on N projectors, each of which is N_F dimensional. The physical motivation is that these are eigenspaces of a Hamiltonian, i.e., projectors to subspaces spanned by degenerate energy eigenstates [35].

9.1.1.1 Wilson Loop

Consider a loop over $N \geq 3$ sets of states from the Hilbert space, each set consisting of $N_F \in \mathbb{N}$ orthonormal states, $\{|u_n(k)\rangle \,|n = 1, \ldots, N_F\}$, with $k = 1, \ldots, N$. We quantify the overlap between set k and set l by the $N_F \times N_F$ *overlap matrix* $M^{(kl)}$, with elements

$$M_{nm}^{(kl)} = \langle u_n(k) \mid u_m(l) \rangle, \qquad (9.1)$$

with $n, m = 1 \ldots, N_F$.

The *Wilson loop* is the product of the overlap matrices along the loop,

$$W = M^{(12)} M^{(23)} \ldots M^{(N-1,N)} M^{(N1)}. \qquad (9.2)$$

We will be interested in the eigenvalues λ_n of the Wilson loop, with $n = 1, \ldots, N_F$,

$$W \underline{v}_n = \lambda_n \underline{v}_n, \tag{9.3}$$

where $\underline{v}_n$ is the nth eigenvector, with $n = 1, \ldots, N_F$.

Note that we could have started the Wilson loop, Eq. (9.2), at the kth group instead of the first one,

$$W^{(k)} = M^{(k,k+1)} M^{(k+1,k+2)} \ldots M^{(N,1)} M^{(1,2)} \ldots M^{(k-1,k)}. \tag{9.4}$$

Although the elements of the matrix $W^{(k)}$ depend on the starting point k, the eigenvalues λ_n do not. Multiplying Eq. (9.3) from the left by $M^{(k,k+1)} \ldots M^{(N,1)}$, we obtain

$$W^{(k)} \left(M^{(k,k+1)} \ldots M^{(N,1)} \right) \underline{v}_n = \lambda_n \left(M^{(k,k+1)} \ldots M^{(N,1)} \right) \underline{v}_n. \tag{9.5}$$

9.1.1.2 $U(N_F)$ Gauge Invariance of the Wilson Loop

We will now show that the eigenvalues of the Wilson loop over groups of Hilbert space vectors only depend on the linear spaces spanned by the vectors of each group. Each group can undergo an independent unitary operation to redefine the vectors, this cannot affect the eigenvalues of the Wilson loop. This is known as the invariance under a $U(N_F)$ gauge transformation.

A simple route to prove the $U(N_F)$ gauge invariance is via the operator $\hat{W}$ defined by the Wilson loop matrix W in the basis of group 1,

$$\hat{W} = \sum_{n=1}^{N_F} \sum_{m=1}^{N_F} |u_n(1)\rangle \, W_{nm} \, \langle u_m(1)| \,. \tag{9.6}$$

This operator has the same eigenvalues as the Wilson loop matrix itself. It can be expressed using the projectors to the subspaces spanned by the groups of states,

$$\hat{P}_k = \sum_n |u_n(k)\rangle \langle u_n(k)| \,. \tag{9.7}$$

The Wilson loop operator reads,

$$\hat{W} = \hat{P}_1 \hat{P}_2 \hat{P}_3 \ldots \hat{P}_N \hat{P}_1. \tag{9.8}$$

We show this explicitly for $N = 3$,

$$\hat{W} = \sum_{n=1}^{N_F} \sum_{n_2=1}^{N_F} \sum_{n_3=1}^{N_F} \sum_{m=1}^{N_F} |u_n(1)\rangle$$

$$\langle u_n(1) \mid u_{n_2}(2)\rangle \langle u_{n_2}(2) \mid u_{n_3}(3)\rangle \langle u_{n_N}(N) \mid u_m(1)\rangle \langle u_m(1)| = \hat{P}_1 \hat{P}_2 \hat{P}_3 \hat{P}_1. \qquad (9.9)$$

The generalization to arbitrary $N \geq 3$ is straightforward.

Equation (9.8) makes it explicit that the Wilson loop operator, and hence, the eigenvalues of the Wilson loop, are $U(N_F)$ gauge invariant.

9.1.2 Wannier States for Degenerate Multiband One-Dimensional Insulators

We now generalize the Wannier states of Sect. 3.2 to a one-dimensional insulator with N_F occupied bands. In case of nondegenerate bands, a simple way to go would be to define a set of Wannier states for each band separately. However, time reversal symmetry forces degeneracies in the bands, at least at time reversal invariant momenta, and so this is not possible. Moreover, even in the nondegenerate case it could be advantageous to mix states from different bands to create more tightly localized Wannier states.

To be specific, and to obtain efficient numerical protocols, we take a finite sample of a one-dimensional insulator of $N = 2M$ unit cells, with periodic boundary conditions. An orthonormal set of negative energy bulk eigenstates reads

$$|\Psi_n(k)\rangle = |k\rangle \otimes |u_n(k)\rangle = \frac{1}{\sqrt{N}} \sum_{m=1}^{N} e^{imk} |m\rangle \otimes |u_n(k)\rangle \qquad (9.10)$$

with, as before, $k \in \{\delta_k, 2\delta_k, \ldots, N\delta_k\}$, and $\delta_k = 2\pi/N$. The index n labels the eigenstates, with $n = 1, \ldots, N_F$ for occupied, negative energy states. The $|u_n(k)\rangle$ are the negative energy eigenstates of the bulk momentum-space Hamiltonian $\hat{H}(k)$. We will not be interested in the positive energy eigenstates.

Although we took a specific set of energy eigenstates above, because of degeneracies at the time-reversal invariant momenta, we really only care about the projector $\hat{P}$ to the negative energy subspace. This is defined as

$$\hat{P} = \sum_{k} \sum_{n=1}^{N_F} |\Psi_n(k)\rangle \langle \Psi_n(k)| = \sum_{k} |k\rangle \langle k| \otimes \hat{P}(k); \qquad (9.11)$$

$$\hat{P}(k) = \sum_{n=1}^{N_F} |u_n(k)\rangle \langle u_n(k)| . \qquad (9.12)$$

9.1.2.1 Defining Properties of Wannier States

We will need a total number $N_F N$ of Wannier states to span the occupied subspace, $|w_n(j)\rangle$, with $j = 1, \ldots, N$, and $n = 1, \ldots, N_F$. These are defined by the usual properties:

$$\langle w_{n'}(j') \mid w_n(j)\rangle = \delta_{j'j}\delta_{n'n} \qquad \text{Orthonormal set} \qquad (9.13a)$$

$$\sum_{j=1}^{N}\sum_{n=1}^{N_F} |w_n(j)\rangle \langle w_n(j)| = \hat{P} \qquad \text{Span occupied subspace}$$

$$(9.13b)$$

$$\forall m: \langle m+1 \mid w_n(j+1)\rangle = \langle m \mid w_n(j)\rangle \qquad \text{Related by translation} \qquad (9.13c)$$

$$\lim_{N\to\infty} \langle w_n(N/2)| (\hat{x} - N/2)^2 |w_n(N/2)\rangle < \infty \quad \text{Localization} \qquad (9.13d)$$

with the addition in Eq. (9.13c) defined modulo N.

The Ansatz of Sect. 3.2 for the Wannier states, Eq. (3.11), generalizes to the multiband case as

$$|w_n(j)\rangle = \frac{1}{\sqrt{N}} \sum_{k=\delta_k}^{N\delta_k} e^{-ijk} \sum_{p=1}^{N_F} U_{np}(k) \, |\Psi_p(k)\rangle. \qquad (9.14)$$

Thus, each Wannier state can contain contributions from all of the occupied bands, the corresponding weights given by a k-dependent unitary matrix $U(k)$.

9.1.2.2 The Projected Unitary Position Operator

As we did in Sect. 3.2, we will specify the set of Wannier states as the eigenstates of the unitary position operator restricted the occupied bands,

$$\hat{X}_P = \hat{P}e^{i\delta_k\hat{x}}\hat{P}. \qquad (9.15)$$

To obtain the Wannier states, we go through the same steps as in Sect. 3.2, with an extra index n. We outline the derivations and detail some of the steps below. You can then check whether the properties required of Wannier states, Eq. (9.13), are fulfilled, in the same way as in the single-band case.

We note that for finite N, the projected unitary position $\hat{X}_P$ is not a normal operator, i.e., it does not commute with its adjoint. As a result, its eigenstates form an orthonormal set only in the thermodynamic limit of $N \to \infty$. Just as in the single-band case, this can be seen as a discretization error, which disappears in the limit $N \to \infty$.

The first step is to rewrite the operator $\hat{X}_P$. For this, consider

$$\left\langle \Psi_{n'}(k') \middle| \hat{X} \middle| \Psi_n(k) \right\rangle = \delta_{k+\delta_k,k'} \left\langle u_{n'}(k+\delta_k) \mid u_n(k) \right\rangle \tag{9.16}$$

where $\delta_{k+\delta_k,k'} = 1$ if $k' = k + \delta_k$, and 0 otherwise. Using this, the projected unitary position operator can be rewritten as

$$\hat{X}_P = \sum_{k'k} \sum_{n',n=1}^{N_F} \left| \Psi_{n'}(k') \right\rangle \left\langle \Psi_{n'}(k') \middle| \hat{X} \middle| \Psi_n(k) \right\rangle \left\langle \Psi_n(k) \right|$$

$$= \sum_{k} \sum_{n',n=1}^{N_F} \left\langle u_{n'}(k+\delta_k) \mid u_n(k) \right\rangle \cdot \left| \Psi_{n'}(k+\delta_k) \right\rangle \left\langle \Psi_n(k) \right|. \tag{9.17}$$

9.1.2.3   Spectrum of the Projected Unitary Position Operator and the Wilson Loop

As in the single-band case, the next step is to consider $\hat{X}_P$ raised to the Nth power. This time, it will not be simply proportional to the projector $\hat{P}$, however. Bearing in mind the orthonormality of the energy eigenstates, $\langle \Psi_n(k) \mid \Psi_{n'}(k') \rangle = \delta_{k'k}\delta_{n'n}$, we find

$$\left(\hat{X}_P \right)^N = \sum_{k} \sum_{mn} W_{mn}^{(k)} \left| \Psi_m(k) \right\rangle \left\langle \Psi_n(k) \right|. \tag{9.18}$$

The Wilson loop matrices $W^{(k)}$, as per Eq. (9.4), are all unitary equivalent, and have the same set of complex eigenvalues,

$$\lambda_n = |\lambda_n| e^{i\theta_n} \quad \text{with} \quad n = 1, \ldots, N_F, \tag{9.19}$$

$$|\lambda_n| \leq 1, \quad \theta_n \in [-\pi, \pi). \tag{9.20}$$

The spectrum of eigenvalues of $\hat{X}_P$ is therefore composed of the Nth roots of these eigenvalues, for $j = 1, \ldots, N$, and $n = 1, \ldots, N_F$,

$$\lambda_{n,j} = e^{i\theta_n/N + ij\delta_k + \log(|\lambda_n|)/N}, \quad \Longrightarrow \quad (\lambda_{n,j})^N = \lambda_n. \tag{9.21}$$

9.1.2.4   Wannier Centers Identified Through the Eigenvalues of the Wilson Loop

As in the single-band case, Sect. 3.2, we identify the phases of the eigenvalues $\lambda_{n,j}$ of the projected position operator $\hat{X}_P$ with the centers of the Wannier states. There are N_F sets of Wannier states, each set containing states that are spaced by distances of 1,

$$\langle x \rangle_{n,j} = \frac{N}{2\pi} \arg \lambda_{n,j} = \langle x \rangle_n + j; \qquad (9.22)$$

$$\langle x \rangle_n = \frac{\theta_n}{2\pi}. \qquad (9.23)$$

The phases θ_n of the N_F eigenvalues of the Wilson loop W are thus identified with the Wannier centers, more precisely, with the amount by which the N_F sets are displaced from the integer positions.

9.2   Time-Reversal Restrictions on Wannier Centers

We will now apply the prescription for Wannier states above to the one-dimensional insulators obtained as slices of a two-dimensional $\hat{\mathcal{T}}^2 = -1$ time-reversal invariant insulator at constant k_y. We will use the language of dimensional reduction, i.e., talk of the bulk Hamiltonian $\hat{H}(k_x, k_y)$ as describing an adiabatic particle pump with k_y playing the role of time. We will use the Wannier center flow, i.e., the quantities $\langle x \rangle_n = \theta_n(k_y)/(2\pi)$ to track the motion of the particles in the bulk during a fictitious pump cycle, $k_y = -\pi \to \pi$.

Time-reversal symmetry places constraints on the Wannier center flow in two ways: it enforces $k_y \leftrightarrow -k_y$ symmetry, i.e., $\theta_n(k_y) = \theta_{n'}(-k_y)$, and it ensures that for $k_y = 0$ and for $k_y = \pi$, the θ_n are doubly degenerate. In this Section we see how these constraints arise.

9.2.1   Eigenstates at k and −k Are Related

A consequence of time-reversal symmetry is that energy eigenstates at $\mathbf{k}$ can be transformed to eigenstates at $-\mathbf{k}$. One might think that because of time-reversal symmetry, energy eigenstates come in time-reversed pairs, i.e., that $\hat{\tau}\hat{H}(-\mathbf{k})^*\hat{\tau}^\dagger = \hat{H}(\mathbf{k})$ would automatically ensure that $|u_n(-\mathbf{k})\rangle = e^{i\phi(\mathbf{k})}\hat{\tau}|u_n(\mathbf{k})^*\rangle$. However, because of possible degeneracies, this is not necessarily the case. The most we can say is that the state $|u_n(-\mathbf{k})\rangle$ is some linear combination of time-reversed eigenstates,

$$|u_n(-\mathbf{k})\rangle = \hat{\tau} \sum_{m=1}^{N_F} (B_{nm}(\mathbf{k}) |u_m(\mathbf{k})\rangle)^* = \sum_{m=1}^{N_F} B_{nm}(\mathbf{k})^* \hat{\tau} |u_m(\mathbf{k})^*\rangle. \qquad (9.24)$$

The coefficients $B_{nm}(\mathbf{k})$ define the unitary *sewing matrix*. An explicit formula for its matrix elements is obtained by multiplying the above equation from the left by $\langle u_a(\mathbf{k})^*| \hat{\tau}^\dagger$, with some $a = 1, \ldots, N_F$. This has the effect on the left- and right-hand side of Eq. (9.24) of

$$\langle u_a(\mathbf{k})^*| \hat{\tau}^\dagger |u_n(-\mathbf{k})\rangle = \left(\langle u_n(-\mathbf{k})| \hat{\tau} |u_a(\mathbf{k})^*\rangle\right)^* ; \qquad (9.25)$$

$$\langle u_a(\mathbf{k})^*| \hat{\tau}^\dagger \sum_{m=1}^{N_F} B_{nm}(\mathbf{k})^* \hat{\tau} |u_m(\mathbf{k})^*\rangle = B_{na}(\mathbf{k})^*, \qquad (9.26)$$

where for the last equation we used the unitarity of $\hat{\tau}$ and the orthonormality of the set $|u_m(\mathbf{k})\rangle$. Comparing the two lines above (and relabeling $a \to m$), we obtain

$$B_{nm}(\mathbf{k}) = \langle u_n(-\mathbf{k})| \hat{\tau} |u_m(\mathbf{k})^*\rangle . \qquad (9.27)$$

Using this definition it is straightforward to show that the sewing matrix is unitary, and that $B_{mn}(-\mathbf{k}) = -B_{nm}(\mathbf{k})$.

9.2.2 Wilson Loops at k_y and $-k_y$ Have the Same Eigenvalues

To see the relation between the Wilson loops at k_y and $-k_y$, we first relate the projectors to the occupied subspace at these momenta. We use a shorthand,

$$\hat{P}_j(k_y) = \begin{cases} \hat{P}(2\pi + j\delta_k, k_y) & \text{if } j \leq 0; \\ \hat{P}(j\delta_k, k_y), & \text{if } j > 0. \end{cases} \qquad (9.28)$$

Using Eq. (9.24), and the unitarity of the sewing matrix B, we find

$$\hat{P}_{-j}(-k_y) = \hat{P}(-\mathbf{k}) = \sum_{n=1}^{N_F} |u_n(-\mathbf{k})\rangle \langle u_n(-\mathbf{k})|$$

$$= \sum_{n=1}^{N_F} \sum_{m=1}^{N_F} \sum_{m'=1}^{N_F} B_{nm}(\mathbf{k})^* \hat{\tau} |u_m(\mathbf{k})^*\rangle B_{nm'}(\mathbf{k}) \langle u_{m'}(\mathbf{k})^*| \hat{\tau}^\dagger$$

$$= \hat{\tau}\hat{P}_j(k_y)^* \tau^\dagger = \hat{\tau}\hat{P}_j(k_y)^T \tau^\dagger. \qquad (9.29)$$

The consequence of Eq. (9.29) for the Wilson loop is

$$\hat{W}(-k_y) = \hat{\tau}\hat{W}(k_y)^T \hat{\tau}^\dagger. \qquad (9.30)$$

We write down the proof explicitly for $N = 6$,

$$
\begin{aligned}
\hat{W}(-k_y) &= \hat{P}_3(-k_y)\hat{P}_2(-k_y)\hat{P}_1(-k_y)\hat{P}_0(-k_y)\hat{P}_{-1}(-k_y)\hat{P}_{-2}(-k_y)\hat{P}_3(-k_y) \\
&= \hat{\tau}\hat{P}_3(k_y)^T \hat{P}_{-2}(k_y)^T \hat{P}_{-1}(k_y)^T \hat{P}_0(k_y)^T \hat{P}_1(k_y)^T \hat{P}_2(k_y)^T \hat{P}_3(k_y)^T \hat{\tau}^\dagger \\
&= \hat{\tau}\hat{W}(k_y)^T \hat{\tau}^\dagger, \qquad (9.31)
\end{aligned}
$$

the generalization to arbitrary even N follows the same lines. The set of eigenvalues of $\hat{W}$ is the same as that of its transpose $\hat{W}^T$, as this holds for any matrix. Moreover, the unitary transformation of $\hat{W}^T$ to $\hat{\tau}\hat{W}^T\hat{\tau}^\dagger$ does not change the eigenvalues either. To summarize, we find that the eigenvalues of the Wilson loop at $-k_y$ are the same as of the Wilson loop at k_y,

$$
\theta_n(k_y) = \theta_n(-k_y). \qquad (9.32)
$$

From Eq. (9.32), we have that the Wannier center flow is symmetric around $k_y = 0$, and hence, also symmetric around $k_y = \pi$. This means that it is enough to examine the Wannier centers from $k_y = 0$ to $k_y = \pi$.

9.2.3   Wilson Loop Eigenvalues at $k_y = 0$ and $k_y = \pi$ Are Doubly Degenerate

We now concentrate on the two special values of the y wavenumber, $k_y = 0$ and $k_y = \pi$, which are mapped unto themselves by time reversal. The one-dimensional Hamiltonians $\hat{H}_{\text{bulk}}(k_x, 0)$ and $\hat{H}_{\text{bulk}}(k_x, \pi)$ are time-reversal invariant. Therefore, due to the Kramers theorem, each of their eigenstates $|\Psi\rangle$ has a time-reversed partner $\hat{\mathscr{T}}|\Psi\rangle$, the two have the same energy, and are orthogonal, $\langle\Psi|\hat{\mathscr{T}}|\Psi\rangle = 0$.

The eigenvalues of the Wilson loop $\hat{W}$ at $k_y = 0$ and $k_y = \pi$ are doubly degenerate. To show this, take an eigenstate of the Wilson loop, $\hat{W}|\Psi\rangle = \lambda|\Psi\rangle$. Using Eq. (9.31), we find

$$
\lambda|\Psi\rangle = \hat{W}|\Psi\rangle = \tau\tau^\dagger\hat{W}\tau\tau^\dagger|\Psi\rangle = \hat{\tau}\hat{W}^T\hat{\tau}^\dagger|\Psi\rangle; \qquad (9.33)
$$

$$
\hat{W}^\dagger\hat{\tau}|\Psi^*\rangle = \lambda^*\hat{\tau}|\Psi^*\rangle. \qquad (9.34)
$$

We obtained line (9.34) by multiplication from the left by $\hat{\tau}$ and complex conjugation, and using the antisymmetry of $\hat{\tau}$. In the final line, we have obtained that the Wilson loop $\hat{W}$ has a left eigenvector with eigenvalue λ^*. Since this is orthogonal to $|\Psi\rangle$, however, the right eigenvalue λ must be at least twice degenerate.

9.3 Two Types of Wannier Center Flow

We now examine the Wannier center flow, i.e., the functions $\theta_n(k_y)$, in time-reversal invariant two-dimensional insulators with $\hat{\mathcal{T}}^2 = -1$. Due to the restrictions of $k_y \leftrightarrow -k_y$ symmetry and degeneracy at $k_y = 0, \pi$, we will find two classes of Wannier center flow. In the trivial class, the Wannier center flow can be adiabatically (i.e., continuously, while respecting the restrictions) deformed to the trivial case, with $\theta_n = 0$ for every n and every k_y. The topological class is the set of cases where this is not possible.

To have a concrete example at hand, we examine the Wannier center flow for the BHZ model of the previous chapter, Eq. (8.38), in a trivial (a) and in a topological (b) case, and a third, more general topological (c) model. All three cases are covered by a modified BHZ Hamiltonian,

$$\hat{H}(\mathbf{k}) = \hat{s}_0 \otimes [(u + \cos k_x + \cos k_y)\hat{\sigma}_z + \sin k_y \hat{\sigma}_y)] + \hat{s}_z \otimes \sin k_x \hat{\sigma}_x + \hat{s}_x \otimes \hat{C} +$$
$$g\hat{s}_z \otimes \hat{\sigma}_y (\cos k_x + \cos 7k_y - 2). \qquad (9.35)$$

For case (a), we set $g = 0$, take the sublattice potential parameter $u = 2.1$, and coupling operator $\hat{C} = 0.02\hat{\sigma}_y$. This is adiabatically connected to the trivial limit of the BHZ model at $u = +\infty$. In case (b), we set $g = 0$, take sublattice potential parameter $u = 1$, and coupling operator $\hat{C} = 0.3\hat{\sigma}_y$, deep in the topological regime. In case (c) we add the extra term to the BHZ model to have a more generic case, with $g = 0.1$, and use $u = 1$, coupling $\hat{C} = 0.1\hat{\sigma}_y$. The Wannier center flows for the three cases are shown in Fig. 9.1.

We consider how adiabatic deformations of the Hamiltonian can affect the Wannier center flow. Focusing on $k_y = 0 \rightarrow \pi$, the center flow consists of branches $\theta_n(k_y)$, that are continuous functions of k_y, beginning at $\theta_n(0)$ and ending at $\theta_n(\pi)$.

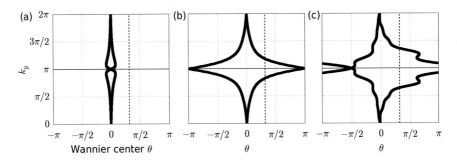

Fig. 9.1 Wannier center flow examples, corresponding to the four-band model of Eq. (9.35). The number of negative-energy, filled bands is $N_F = 2$, hence the graphs show the evolution of two Wannier centers (θ) as a function of the momentum k_y. See text for parameter specifications

Due to an adiabatic deformation,

- A branch can bend while $\theta_n(0)$ and $\theta_n(\pi)$ are fixed;
- The endpoint at $k_y = 0$ (or $k_y = \pi$) of a branch can shift: in that case, the endpoint of the other branch, the Kramers partner at $k_y = 0$ (or $k_y = \pi$) is shifted with it;
- Branches θ_n and θ_m can recombine: a crossing between them at some k_y can turn into an avoided crossing.

Consider the example of Fig. 9.1. Bending of the branches and shifting of the endpoints can bring case (a) to a trivial case, where all branches are vertical, $\theta_n(k_y) = 0$ for every n and k_y. Case (b) can be deformed to case (c). Notice, however, that neither cases (b) nor (c) can be deformed to the trivial case.

9.3.1 Bulk Topological Invariant

We define the bulk topological invariant N_{bulk}, by choosing some fixed $\tilde{\theta} \in [-\pi, \pi)$, and asking for the parity of the number of times the Wannier center flow crosses this $\tilde{\theta}$, as k_y is varied between 0 and π. In formulas,

$$\tilde{\theta} \in [-\pi, \pi); \tag{9.36}$$

$$N_n(\tilde{\theta}) = \text{Number of solutions } k_y \text{ of } \theta_n(k_y) = \tilde{\theta}, \ k_y \in [0, \pi]; \tag{9.37}$$

$$N_{\text{bulk}} = \left(\sum_{n=1}^{N_F} N_n(\tilde{\theta}) \right) \mod 2 \qquad (\text{independent of } \tilde{\theta}). \tag{9.38}$$

The number N_{bulk} is invariant under adiabatic deformations of the bulk Hamiltonian, as can be shown by considering the possible changes. Bending of a branch θ_n can create or destroy solutions of $\theta_n(k_y) = \tilde{\theta}$, but only pairwise. Shifting of the endpoint can create or destroy single solutions of $\theta_n(k_y) = \tilde{\theta}$, but in that case, a single solution of $\theta_m(k_y) = \tilde{\theta}$, is also created/destroyed, where θ_m is the Kramers partner of θ_n at the endpoint. Finally, recombination of branches cannot change the number of crossings. The number N_{bulk} is also invariant under a shift of $\tilde{\theta}$, as already announced. A shifting of $\tilde{\theta}$ is equivalent to a shifting of the Wannier center flow, whose effects we already considered above.

9.3.1.1 The Bulk Topological Invariant Is the $\mathbb{Z}_2$ Invariant of the Previous Chapter

The full proof that the bulk invariant N_{bulk} is the same as the parity D of the number of edge state pairs, Eq. (8.40), is quite involved [16, 37]. We content ourselves with just pointing out here that both N_{bulk} and D represent obstructions to deform the

Hamiltonian adiabatically to the so-called atomic limit, when the unit cells are completely disconnected from each other. Clearly, switching of a charge pump requires that there are no edge states present, and therefore, $N_{\text{bulk}} = 0$ requires $D = 0$. To show that the converse is true is more complicated, and we do not discuss it here.

9.4   The $\mathbb{Z}_2$ Invariant for Systems with Inversion Symmetry

For two-dimensional time-reversal invariant insulators with inversion (i.e., parity) symmetry, the $\mathbb{Z}_2$ topological invariant becomes very straightforward. We state the result below, and leave the proof as an exercise for the reader.

9.4.1   Definition of Inversion Symmetry

As introduced in Sect. 3.3, the operation of inversion, $\hat{\Pi}$, acts on the bulk momentum-space Hamiltonian using an operator $\hat{\pi}$, by

$$\hat{\Pi}\hat{H}(\mathbf{k})\hat{\Pi}^{-1} = \hat{\pi}\hat{H}(-\mathbf{k})\hat{\pi}^{\dagger}. \tag{9.39}$$

We now require the operator $\hat{\pi}$ not only to be independent of the wavenumber $\mathbf{k}$, to be unitary, Hermitian, but also to commute with time reversal, i.e.,

$$\hat{\Pi}^{\dagger}\hat{\Pi} = 1; \qquad \hat{\Pi}^2 = 1; \qquad \hat{\mathscr{T}}\hat{\Pi} = \hat{\Pi}\hat{\mathscr{T}}. \tag{9.40}$$

9.4.2   At a Time-Reversal Invariant Momentum, the Kramers Pairs Have the Same Inversion Eigenvalue

Consider the time-reversal invariant momenta, Γ_j. In the BHZ model, these are $(k_x, k_y) = (0,0), (0, \pi), (\pi, 0), (\pi, \pi)$. In general there are 2^d such momenta in a d-dimensional lattice model. Each eigenstate $|u(\Gamma_j)\rangle$ of the bulk momentum-space Hamiltonian at these momenta has an orthogonal Kramers pair $\hat{\mathscr{T}}|u(\Gamma_j)\rangle$,

$$\hat{H}(\Gamma_j)|u(\Gamma_j)\rangle = E|u(\Gamma_j)\rangle \implies \hat{H}(\Gamma_j)\hat{\mathscr{T}}|u(\Gamma_j)\rangle = E\hat{\mathscr{T}}|u(\Gamma_j)\rangle. \tag{9.41}$$

If $\hat{H}$ is inversion symmetric, $|u\rangle$ can be chosen to be an eigenstate of $\hat{\pi}$ as well, since

$$\hat{\pi}\hat{H}(\Gamma_j)\hat{\pi} = \hat{H}(-\Gamma_j) = \hat{H}(\Gamma_j). \tag{9.42}$$

Therefore,

$$\hat{\pi}\left|u(\Gamma_j)\right\rangle = \pm \left|u(\Gamma_j)\right\rangle. \tag{9.43}$$

The Kramers pair of $\left|u(\Gamma_j)\right\rangle$ has to have the same inversion eigenvalue as $\left|u(\Gamma_j)\right\rangle$,

$$\hat{\pi}\,\hat{\mathscr{T}}\left|u(\Gamma_j)\right\rangle = \hat{\mathscr{T}}\,\hat{\pi}\left|u(\Gamma_j)\right\rangle = \pm\hat{\mathscr{T}}\left|u(\Gamma_j)\right\rangle. \tag{9.44}$$

In a system with both time-reversal and inversion symmetry, we get 2^d topo-logical invariants of the bulk Hamiltonian, one for each time-reversal invariant momentum Γ_j. These are the products of the parity eigenvalues $\xi_m(\Gamma_j)$ of the occupied Kramers pairs at Γ_j. However, inversion symmetry is usually broken at the edges, and so these invariants do not give rise to robust edge states.

The product of the inversion eigenvalues of all occupied Kramers pairs at all the time-reversal invariant momenta Γ_j is the same as the $\mathbb{Z}_2$ invariant,

$$(-1)^{N_{\text{bulk}}} = \prod_j \prod_m \xi_m(\Gamma_j). \tag{9.45}$$

We leave the proof of this useful result as an exercise for the reader.

9.4.3   Example: The BHZ Model

A concrete example for inversion symmetry is given by the BHZ model of Sect. 8.2, with no coupling $\hat{C} = 0$. It can be checked directly that this has inversion symmetry, with

$$\hat{\pi} = \hat{s}_0 \otimes \hat{\sigma}_z. \tag{9.46}$$

To calculate the $\mathbb{Z}_2$ invariant of the BHZ model, we take the four time-reversal invariant momenta, $\mathbf{k}_1, \mathbf{k}_2, \mathbf{k}_3, \mathbf{k}_4$, are the combinations of k_x, k_y with $k_x = 0, \pi$ and $k_y = 0, \pi$. The Hamiltonian $\hat{H}_{\text{BHZ}}(k_x, k_y)$ at these momenta is proportional to the inversion operator,

$$\hat{H}_{\text{BHZ}}(\mathbf{k}_1 = 0, 0) = (u + 2)\hat{\pi}; \qquad \hat{H}_{\text{BHZ}}(\mathbf{k}_4 = \pi, \pi) = (u - 2)\hat{\pi}; \tag{9.47}$$

$$\hat{H}_{\text{BHZ}}(\mathbf{k}_2 = 0, \pi) = u\hat{\pi}; \qquad \hat{H}_{\text{BHZ}}(\mathbf{k}_3 = \pi, 0) = u\hat{\pi}. \tag{9.48}$$

In these cases the Hamiltonian and the inversion operator obviously have the same eigenstates. At each time-reversal invariant momentum, two of these states form one occupied Kramers pair and the two others one empty Kramers pair. If $u > 2$, at all four time-reversal invariant momenta, the occupied Kramers pair is the one with inversion eigenvalue (parity) of -1, and so Eq. (9.45) gives $N_{\text{bulk}} = 0$. Likewise, if

$u < -2$, the eigenvalues are all $+1$, and we again obtain $N_{bulk} = 0$. For $0 < u < 2$, we have P eigenvalues $-1, -1, +1, -1$ at the four time-reversal invariant momenta $\mathbf{k}_1, \mathbf{k}_2, \mathbf{k}_3, \mathbf{k}_4$, respectively, whereas if $-2 < u < 0$, we have $-1, +1, +1, +1$. In both cases, Eq. (9.45) gives $N_{bulk} = 1$. This indeed is the correct result, that we obtained via the Chern number earlier.

Problems

9.1 Inversion symmetry and interlayer coupling in the BHZ model

Consider the BHZ model with layer coupling $\hat{C} = C\hat{s}_x \otimes \hat{\sigma}_y$. This breaks the inversion symmetry $\hat{\pi} = \hat{s}_0 \otimes \hat{\sigma}_z$. Nevertheless, the Wannier centers of the Kramers pairs at $k_y = 0$ and $k_y = \pi$ are stuck to $\theta = 0$ or $\theta = \pi$, and are only shifted by the extra term $\propto \hat{s}_z \hat{\sigma}_y$ added to the BHZ model in Eq. (9.35). Can you explain why? (hint: extra inversion symmetry)

9.2 Proof of the formula for the $\mathbb{Z}_2$ invariant of an inversion-symmetric topological insulator

Show, using the results of Sect. 3.3, that the $\mathbb{Z}_2$ invariant of a two-dimensional time-reversal invariant and inversion symmetric insulator can be expressed using Eq. (9.45).

Chapter 10
Electrical Conduction of Edge States

It is well known that the electrical conduction of ordinary metallic samples at room temperature shows the following two characteristics. First, there is a linear relation between the electric current I that flows through the sample and the voltage V that drops between the two ends of the sample: $I/V = G \equiv R^{-1}$, where G (R) is the conductance (resistance) of the sample. Second, the conductances G_i of different samples made of the same metal but with different geometries show the regularity $G_i L_i / A_i = \sigma$ for $\forall i$, where L_i is the length of the sample and A_i is the area of its cross section. The material-specific quantity σ is called the conductivity. Conductors obeying both of these relations are referred to as Ohmic.

Microscopic theories describing the above behavior (e.g., Drude model, Boltzmann equation) rely on models involving impurities, lattice vibrations, and electron scattering within the material. Electrical conduction in clean (impurity-free) nanostructures at low temperature might therefore qualitatively deviate from the Ohmic case. Here, we demonstrate such deviations on a simple zero-temperature model of a two-dimensional, perfectly clean, constant-cross-section metallic wire, depicted in Fig. 10.1a. Then we describe how scattering at static impurities affects the conduction in general. Finally, we discuss electrical conduction in two-dimensional topological insulators. Even though these materials are band insulators judged from their bulk band structure, in finite-sized samples their edge states do conduct. As a central result in the field of topological insulators, we point out that this conduction shows a strong robustness against impurity scattering.

In previous chapters, and also here, we use two-dimensional models. The real samples used in electronic transport experiments, of course, are three dimensional. However, the electrons participating in transport are often confined to a flat, quasi-two-dimensional spatial region, which allows their description using two-dimensional models. This is the case, for example, in semiconductor quantum wells (see below the example of a HgTe-based quantum well). In such structures, the confinement along the third, say, z, dimension, is very narrow, therefore the

© Springer International Publishing Switzerland 2016
J.K. Asbóth et al., *A Short Course on Topological Insulators*, Lecture Notes in Physics 919, DOI 10.1007/978-3-319-25607-8_10

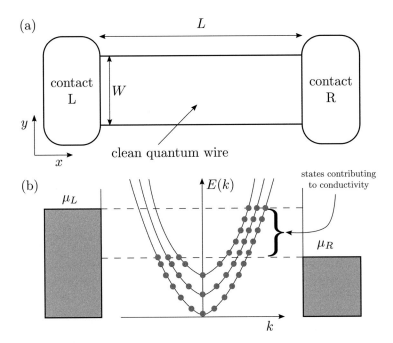

Fig. 10.1 (a) Schematic representation of a clean quantum wire contacted to two electron reservoirs (contacts). (b) Occupations of electronic states in the contacts and the quantum wire in the nonequilibrium situation when a finite voltage V is applied between the left and right reservoirs

excitation energies between the corresponding transverse modes are large, and hence the transitions between these can be disregarded. In that case, one can often focus on a single transverse mode along z, the one closest to the electronic Fermi energy, and disregard all other modes, thereby arriving to a two-dimensional model for the electrons. Alternatively, a few of the transverse modes in the vicinity of the Fermi energy might be relevant; in that case, those modes can still be incorporated in a two-dimensional model in the form of an integral degree of freedom.

10.1   Electrical Conduction in a Clean Quantum Wire

As shown in Fig. 10.1a, take a wire that lies along the x axis, with length L and width W. Its width W along y can be defined by, for example, an electric confinement potential or lattice termination. Each electronic energy eigenfunction $|l, k\rangle$ in such a wire is a product of a standing wave along y, labeled by a positive integer l, and a plane wave propagating along x, labeled by a real wave number k (see Eq. (8.41)). A typical set of dispersion relations E_{lk} ('subbands') for three different l indices is shown in Fig. 10.1b.

We also make assumptions on the two metallic contacts that serve as source and drain of electrons. We assume that the electrons in each contact are in thermal equilibrium, but the Fermi energies in the contacts differ by $\mu_L - \mu_R = |e|V > 0$. (Note that in this chapter, proper physical units are used, hence constants such as the elementary charge $|e|$, reduced Planck's constant $\hbar$, lattice constant a are reinstated.) We consider the *linear conductance*, that is, the case of an infinitesimal voltage $|e|V \to 0$. We further assume that both contacts absorb every incident electron with unit probability, and that the energy distribution of the electrons they emit is the thermal distribution with the respective Fermi energy.

These assumptions guarantee that the right-moving (left-moving) electronic states in the wire are occupied according to the thermal distribution of the left (right) contact, as illustrated in Fig. 10.1b. Now, we work with electron states normalized to the area of the channel. It is a simple fact that with this normalization convention, a single occupied state in the lth channel, with wave number k carries an electric current of $\frac{-|e|v_{lk}}{N_x a}$, where $N_x a$ is the length of the wire, and $v_{lk} = \frac{1}{\hbar}\frac{dE_{lk}}{dk}$ is the group velocity of the considered state. Therefore, the current flowing through the wire is

$$I = -|e|\frac{1}{N_x a}\sum_{lk} v_{lk}\left[f(E_{lk} - \mu_L) - f(E_{lk} - \mu_R)\right], \tag{10.1}$$

where $f(\epsilon) = \left(\exp\frac{\epsilon}{k_B T} + 1\right)^{-1}$ is the Fermi-Dirac distribution. Converting the k sum to an integral via $\frac{1}{N_x a}\sum_k \ldots \mapsto \int_{-\pi/a}^{\pi/a}\frac{dk}{2\pi}\ldots$ yields

$$I = -|e|\sum_n \int_{-\pi/a}^{\pi/a}\frac{dk}{2\pi}\frac{1}{\hbar}\frac{dE_{lk}}{dk}\left[f(E_{lk} - \mu_L) - f(E_{lk} - \mu_R)\right]. \tag{10.2}$$

The Fermi-Dirac distribution has a sharp edge at zero temperature, implying

$$I = -\frac{|e|}{h}M\int_{\mu_R}^{\mu_L} dE = -\frac{|e|}{h}(\mu_L - \mu_R)M = M\frac{e^2}{h}V \tag{10.3}$$

Note that the first equality in (10.3) holds only if the number of subbands intersected by μ_L and μ_R are the same, which is indeed the case if the voltage V is small enough. The number of these subbands, also called 'open channels', is denoted by the integer M. From (10.3) it follows that the conductance of the wire is an integer multiple of e^2/h (commonly referred to as 'quantized conductance'):

$$G = \frac{e^2}{h}M. \tag{10.4}$$

The numerical value of e^2/h is approximately 40 µS (microsiemens), which corresponds to a resistance of approximately 26 kΩ. Note that the conductance quantum is defined as $G_0 = 2e^2/h$, i.e., as the conductance of a single open channel with twofold spin degeneracy.

It is instructive to compare the conduction in our clean quantum wire to the ordinary Ohmic conduction summarized above. According to (10.3), the proportionality between voltage and current holds for a clean quantum wire as well as for an ordinary metal. However, the dependence of the conductance on the length of the sample differs qualitatively in the two cases: in an ordinary metal, a twofold increase in the length of the wire halves the conductance, whereas the conductance of a clean quantum wire is insensitive to length variations.

Whether the conductance of the clean quantum wire is sensitive to variations of the wire width depends on the nature of the transversal modes. Conventional quantum wires that are created by a transverse confinement potential have a subband dispersion similar to that in Fig. 10.1b. There, the energy separation between the subbands decreases as the width of the wire is increased, therefore the number of subbands available for conduction increases. This leads to an increased conductance for an increased width, similarly to the case of ordinary metals. If, however, we consider a topological insulator, where the current is carried by states localized to the edges of the wire, the conductance of the wire will be insensitive to the width of the wire.

10.2   Phase-Coherent Electrical Conduction in the Presence of Scatterers

Having calculated the conductance (10.4) of a clean quantum wire, we now describe how this conductance is changed by the presence of impurities. We analyze the model shown in Fig. 10.2, where the disordered region, described by a scattering matrix S, is connected to the two contacts by two identical clean quantum wires, also called 'leads' in this context.

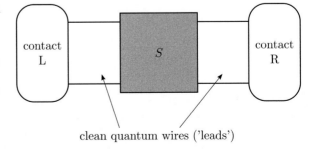

Fig. 10.2 Simple model of a phase-coherent conductor in the presence of scatterers. The ideal contacts L and R are connected via ideal leads to the disordered region represented by the scattering matrix S

First, we consider the case when each lead supports a single open channel. The current in the lead connecting contact L and the scattering region consists of a contribution from right-moving states arriving from contact L and partially backscattered with probability $R = |r|^2$, and from left-moving states arriving from contact R and partially transmitted with probability $T' = |t'|^2$:

$$I = -|e|\frac{1}{N_x a} \sum_k v_k \left[(1 - R(E_k))f_L(E_k) - T'(E_k)f_R(E_k)\right] \qquad (10.5)$$

Converting the k sum to an integral, assuming that the transmission and reflection probabilities are independent of energy in the small energy window between μ_R and μ_L, and using $1 - R = T = T'$, we arrive at

$$I = -\frac{|e|}{h}T \int_{\mu_R}^{\mu_L} dE \left[f_L(E) - f_R(E)\right] = \frac{e^2}{h}TV, \qquad (10.6)$$

which implies that the conductance can be expressed through the transmission coefficient T:

$$G = \frac{e^2}{h}T. \qquad (10.7)$$

The result (10.7) can be straightforwardly generalized to the case when the leads support more than one open channel. The generalized result for the conductance, also known as the Landauer formula, reads:

$$G = \frac{e^2}{h}\sum_{n=1}^{M} T_n, \qquad (10.8)$$

where T_n are the transmission eigenvalues of the scattering matrix i.e., the real eigenvalues of the Hermitian matrix $tt^\dagger$, as defined in the preceding chapter.

10.3   Electrical Conduction in Two-Dimensional Topological Insulators

After presenting the Landauer formula as a generic tool to describe electrical conduction of a phase-coherent metal, we will use it know to characterize the conductances of various two-dimensional topological insulator samples.

10.3.1   Chern Insulators

In Sect. 6.2, we have seen that an impurity-free straight strip of a topologically nontrivial Chern insulator supports edge states. The relation between the Chern number Q of the Chern insulator and the numbers of edge states at a single edge at a given energy E, propagating 'clockwise' ($N_+(E)$) and 'anticlockwise' ($N_-(E)$), is $Q = N_+(E) - N_-(E)$. In addition, in Sect. 6.3 it was shown that any segment of the edge of a disordered Chern insulator with Chern number Q and an arbitrary geometry supports $|Q|$ chiral edge modes. Here we show that existence of these edge modes leads to experimentally detectable effects in the electrical transport through Chern insulator samples.

We consider a transport setup where the Chern insulator is contacted with two metallic electrodes, as shown in Fig. 10.3. In this discussion, we rely on the usual assumptions behind the Landauer formula: phase-coherence of the electrons,

Fig. 10.3 A disordered sample of Chern insulator, with contacts 1 and 2, that can be used to pass current through the sample in order to detect edge states

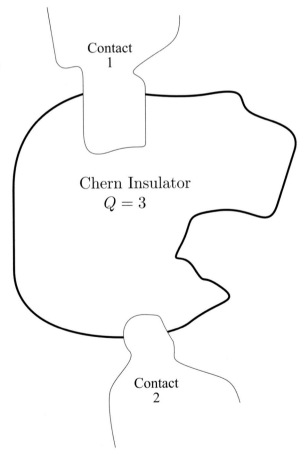

good contact between contacts and sample, and large spatial separation of the two electrodes ensuring the absence of tunneling contributions to the conductance.

In the following list, we summarize how the phase-coherent electrical conductance of a Chern insulator varies with the sample geometry, absence or presence of disorder, and the value of the electronic Fermi energy.

1. Disorder-free sample with a strip geometry (see Fig. 10.1a)

   a. *Fermi energy lies in a band.* In this case, the sample is a clean quantum wire (see Sect. 10.1) with an integer number of open channels. The corresponding transversal wave functions might or might not be localized to the sample edges, and therefore the number of channels might be different from any combination of Q, N_+ or N_-. According to Eq. (10.4), the conductance of such a clean quantum wire is quantized and insensitive to the length of the sample. Furthermore, the conductance grows in a step-like fashion if the width of the sample is increased.

   b. *Fermi energy lies in the gap.* The sample is a clean quantum wire with open channels that are all localized to the sample edges. The number of those channels is $N_+(E) + N_-(E)$, where E is the Fermi energy: in each of the two possible direction of current flow, there are N_+ channels on one edge and N_- on the other edge that contribute to conduction. Conductance is finite and quantized, a behavior rather unexpected from an insulator. The conduction is not Ohmic, as the conductance is insensitive to both the length and the width of the sample.

2. Disordered sample with an irregular shape (see Fig. 10.3):

   a. *Fermi energy lies in a band.* Because of the presence of disorder, the electrical conduction of such a sample might be Ohmic. There are no protected edge states at the Fermi energy.

   b. *Fermi energy lies well within the gap.* According to Sect. 6.3, any edge segment of such a sample supports Q reflectionless chiral edge modes at the Fermi energy. Therefore, conductance is typically quantized, $G = |Q|e^2/h$, although, disorder permitting, it might in principle be larger than this value. The quantized conductance is insensitive to changes in the geometry or the disorder configuration. This transport property, unexpected for an insulator, let alone for one with disorder, is a hallmark of Chern insulators.

In the case of two-dimensional samples there is often an experimental possibility of tuning the electronic Fermi energy *in situ* by controlling the voltage applied between the sample and a nearby metallic plate (*gate electrode*), as discussed in Sect. 10.4. This allows, in principle, to observe the changes in the electrical conduction of the sample as the Fermi energy is tuned across the gap.

10.3.2 Two-Dimensional Time-Reversal Invariant Topological Insulators with $\hat{\mathscr{T}}^2 = -1$

In the following list, we summarize the predictions of the Landauer formalism for the conductance of two-dimensional time-reversal invariant topological insulators with $\hat{\mathscr{T}}^2 = -1$ ('$D = 1$ insulators' for short).

1. Disorder-free sample with a strip geometry:

   a. *Fermi energy lies in a band.* A simple consequence of the Landauer formula is that phase-coherent conductance of an impurity-free $D = 1$ topological insulator of the strip geometry shown in Fig. 10.1a is quantized. The conductance grows if the width of the strip is increased, but it is insensitive to changes in the length.
   b. *Fermi energy lies in the gap.* Only edge channels are open in this case. These also provide conductance quantization. As the number of edge-state Kramers pair per edge is odd, the conductance might be $2e^2/h$, $6e^2/h$, $10e^2/h$, etc. Conductance is insensitive to width or length changes of the sample.

2. Disordered sample with an irregular shape and time-reversal symmetric disorder:

   a. *Fermi energy lies in a band.* The electrical conduction might be Ohmic.
   b. *Fermi energy lies in the gap.* We have shown in Chap. 8 that a $D = 1$ insulator supports one protected edge-state Kramers pair per edge, which allows for reflectionless electronic transmission if only time-reversal symmetric defects are present. The Landauer formula (10.8) implies, for typical cases, $G = 2e^2/h$ for such a sample, as one edge state per edge contributes to conduction. The conductance might also be larger, provided that the number of edge-state Kramers pairs is larger than 1 and disorder is ineffective in reducing the transmission of the topologically unprotected pairs.

We note that in real materials with $\mathbb{Z}_2$ invariant $D = 1$, various mechanisms might lead to backscattering and, in turn, to $G < 2e^2/h$. Examples include time-reversal-symmetry-breaking impurities, time-reversal symmetric impurities that bridge the spatial distance between the edges (see Chap. 8), hybridization of edge states from opposite edges in narrow samples, and inelastic scattering on phonons or spinful impurities.

10.4 An Experiment with HgTe Quantum Wells

Electrical transport measurements [21] on appropriately designed layers of the semiconductor material mercury-telluride (HgTe) show signatures of edge-state conduction in the absence of magnetic field. These measurements are in line with the

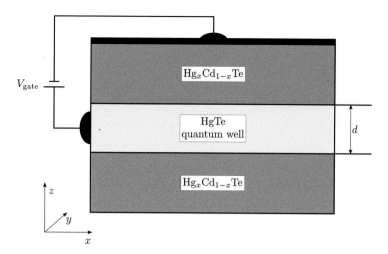

Fig. 10.4 Schematic representation of a HgTe quantum well of width d, sandwiched between two $Hg_xCd_{1-x}Te$ layers. Electrons are confined to the HgTe layer, and their Fermi energy can be tuned *in situ* by adjusting the voltage V_{gate} of the metallic electrode on the top of the sample (black). For a more accurate description of the experimental arrangement, see [20]

theoretical prediction that a HgTe layer with a carefully chosen thickness can realize a topologically nontrivial $(D = 1)$ two-dimensional time-reversal invariant insulator with $\hat{\mathcal{T}}^2 = -1$. In this section, we outline the main findings of this experiment, as well as its relation to the BHZ model introduced and discussed in Chap. 8.

The experiments are performed on sandwich-like structures formed by a few-nanometer thick HgTe layer (*quantum well*) embedded between two similar layers of the alloy $Hg_xCd_{1-x}Te$, as shown in Fig. 10.4. (In the experiment reported in [21], the alloy composition $x = 0.3$ was used.) In this structure, the electronic states with energies close the Fermi energy are confined to the HgTe layer that is parallel to the x-y plane in Fig. 10.4. The energy corresponding to the confinement direction z is quantized. The carriers are free to move along the HgTe layer, i.e., parallel to the x-y plane, therefore two-dimensional subbands are formed in the HgTe quantum well. Detailed band structure calculations of [5] show that as the the thickness d of the HgTe layer is decreased, the lowermost conduction subband and the uppermost valence subband touch at a critical thickness $d = d_c$, and the gap is reopened for even thinner HgTe layers. (For the alloy composition $x = 0.3$ used in the experiment, the critical thickness is $d_c \approx 6.35$ nm.) This behavior is illustrated schematically in Fig. 10.5, which illustrates the electronic dispersions of the uppermost valence subband and the the lowermost conduction subband, at the center of the Brillouin zone, for three different thicknesses of the quantum well.

Band structure calculations have also revealed a connection between the subbands depicted in Fig. 10.5 and the BHZ model introduced and discussed in Chap. 8. The 4×4 effective Hamiltonian describing the two spinful two-dimensional subbands around their extremum point at the centre of the HgTe Brillouin zone

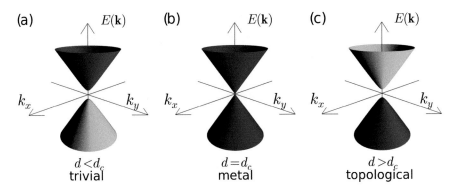

Fig. 10.5 Evolution of the two-dimensional band structure of a HgTe quantum well as a function of its thickness d. (**a**) For a thin quantum well below the critical thickness $d < d_c$, the band structure has a gap and the system is a trivial insulator. (**b**) At a critical thickness $d = d_c$, the band gap closes and the system is metallic. (**c**) For a thick quantum well with $d > d_c$, the band gap reopens and the system becomes a two-dimensional topological insulator

resembles the low-energy continuum Hamiltonian derived from the BHZ lattice model in the vicinity of the $u \approx -2$ value. Changing the thickness d of the HgTe layer corresponds to a change in the parameter u of the BHZ model, and the critical thickness $d = d_c$ corresponds to $u = -2$ and, consequently, a zero mass in the corresponding two-dimensional Dirac equation.

As a consequence of the strong analogy of the band structure of the HgTe quantum well and that of the BHZ model, it is expected that either for $d < d_c$ or for $d > d_c$ the material is a $D = 1$ insulator with a single Kramers pair of edge states. Arguments presented in [5] suggest that the thick quantum wells with $d > d_c$ are topologically nontrivial.

Electrical transport measurements were carried out in HgTe quantum wells patterned in the Hall bar geometry shown in Fig. 10.6. The quantity that has been used in this experiment to reveal edge-state transport is the four-terminal resistance $R_{14,23} = V_{23}/I_{14}$, where V_{23} is the voltage between contacts 2 and 3, and I_{14} is the current flowing between contacts 1 and 4. This quantity $R_{14,23}$ was measured for various devices with different thicknesses d, below and above the critical thickness d_c, of the HgTe layer, and for different values of the Fermi energy. The latter can be tuned *in situ* by controlling the voltage between the HgTe layer and a metallic 'gate' electrode on the top of the layered semiconductor structure, as shown in Fig. 10.4.

To appreciate the experimental result, let us first derive the four-terminal resistance $R_{14,23}$ for such a device. To this end, we express V_{23} with I_{14}. Ohm's law implies $V_{23} = I_{23}/G_{23}$, where I_{23} is the current flowing through the edge segment between contacts 2 and 3, whereas G_{23} is the conductance of that edge segment. Furthermore, as the current I_{14} flowing through terminals 1 and 4 is equally divided between the upper and lower edges, the relation $I_{23} = I_{14}/2$ holds, implying the result $R_{14,23} = 1/(2G_{23})$.

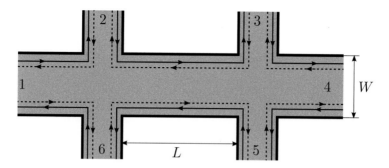

Fig. 10.6 HgTe quantum well patterned in the Hall-bar geometry (gray area). Numbered terminals lead to metallic contacts. Solid and dashed lines depict counterpropagating edge states

If the Fermi energy lies in the bands neighboring the gap, then irrespective of the topological invariant of the system, the HgTe quantum well behaves as a good conductor with $G_{23} \gg e^2/h$, implying $R_{14,23} \ll h/e^2$. If the Fermi energy is tuned to the gap in the topologically nontrivial case $d > d_c$, then $G_{23} = e^2/h$ and therefore $R_{14,23} = h/(2e^2)$. This holds, of course, only at a temperature low enough and a sample size small enough such that phase coherence is guaranteed. The presence of static time-reversal invariant defects is included. If the system is topologically trivial ($d < d_c$), then there is no edge transport, and the quantum well is a good insulator with $R_{14,23} \gg h/e^2$.

The findings of the experiments are consistent with the above expectations. Furthermore, the four-terminal resistance of topologically nontrivial HgTe layers with different widths were measured, with the resistance found to be approximately constant function of the width W of the Hall bar. This is a further indication that the current in these samples is carried by edge states.

To wrap up this chapter, we note that InAs/GaSb bilayer quantum wells are an alternative semiconductor material system where two-dimensional topological insulators can be realized [9, 22]. Graphene is believed to be a two-dimensional topological insulator as well [19], even though its energy gap between the valence and conduction band, induced by spin-orbit interaction and estimated to be of the order of μeV, seems to be too small to allow for the detection of edge-state transport even at the lowest available temperatures. The concept of a time-reversal invariant topological insulator can be extended to three-dimensional crystals as well, where the role of the edge states is played by states localized to the two-dimensional surface of the three-dimensional material. The description of such systems is out of the scope of the present course; the interested reader might consult, e.g., [4, 17].

References

1. A.A. Aligia, G. Ortiz, Quantum mechanical position operator and localization in extended systems. Phys. Rev. Lett. **82**, 2560–2563 (1999)
2. Y. Ando, Topological insulator materials. J. Phys. Soc. Jpn. **82**(10), 102001 (2013)
3. G. Bastard, *Wave Mechanics Applied to Semiconductor Heterostructures* (Les Editions de Physique, Les Ulis, 1988)
4. B.A. Bernevig, *Topological Insulators and Topological Superconductors* (Princeton University Press, Princeton, 2013)
5. B.A. Bernevig, T.L. Hughes, S.-C. Zhang, Quantum spin hall effect and topological phase transition in HgTe quantum wells. Science **314**, 1757 (2006)
6. M.V. Berry, Quantal phase factors accompanying adiabatic changes. Proc. R. Soc. Lond. A **392**, 45–57 (1984)
7. J.C. Budich, B. Trauzettel, From the adiabatic theorem of quantum mechanics to topological states of matter. Physica Status Solidi RRL **7**(1–2), 109–129 (2013)
8. C.-Z. Chang, J. Zhang, X. Feng, J. Shen, Z. Zhang, M. Guo, K. Li, Y. Ou, P. Wei, L.-L. Wang, et al., Experimental observation of the quantum anomalous hall effect in a magnetic topological insulator. Science **340**(6129), 167–170 (2013)
9. L. Du, I. Knez, G. Sullivan, R.-R. Du, Robust helical edge transport in gated InAs/GaSb bilayers. Phys. Rev. Lett. **114**, 096802 (2015)
10. M. Franz, L. Molenkamp, *Topological Insulators*, vol. 6 (Elsevier, Oxford, 2013)
11. L. Fu, C.L. Kane, Time reversal polarization and a Z_2 adiabatic spin pump. Phys. Rev. B **74**, 195312 (2006)
12. T. Fukui, Y. Hatsugai, H. Suzuki, Chern numbers in discretized brillouin zone: efficient method of computing (spin) hall conductances. J. Phys. Soc. Jpn. **74**(6), 1674–1677 (2005)
13. I.C. Fulga, F. Hassler, A.R. Akhmerov, Scattering theory of topological insulators and superconductors. Phys. Rev. B **85**, 165409 (2012)
14. A. Garg, Berry phases near degeneracies: Beyond the simplest case. Am. J. Phys. **78**(7), 661–670 (2010)
15. D.J. Griffiths, *Introduction to Quantum Mechanics* (Pearson Education Limited, Harlow, 2014)
16. F. Grusdt, D. Abanin, E. Demler, Measuring Z_2 topological invariants in optical lattices using interferometry. Phys. Rev. A **89**, 043621 (2014)
17. M.Z. Hasan, C.L. Kane, Colloquium: topological insulators. Rev. Mod. Phys. **82**, 3045 (2010)
18. B.R. Holstein, The adiabatic theorem and Berry's phase. Am. J. Phys. **57**(12), 1079–1084 (1989)
19. C.L. Kane, E.J. Mele, Z_2 topological order and the quantum spin hall effect. Phys. Rev. Lett. **95**, 146802 (2005)

© Springer International Publishing Switzerland 2016
J.K. Asbóth et al., *A Short Course on Topological Insulators*, Lecture Notes in Physics 919, DOI 10.1007/978-3-319-25607-8

20. M. König, H. Buhmann, L.W. Molenkamp, T. Hughes, C.-X. Liu, X.-L. Qi, S.-C. Zhang, The quantum spin hall effect: theory and experiment. J. Phys. Soc. Jpn. **77**(3), 031007 (2008)
21. M. König, S. Wiedmann, C. Brüne, A. Roth, H. Buhmann, L.W. Molenkamp, X.-L. Xi, S.-C. Zhang, Quantum spin hall insulator state in HgTe quantum wells. Science **318**(6), 766–770 (2007)
22. C. Liu, T.L. Hughes, X.-L. Qi, K. Wang, S.-C. Zhang, Quantum spin hall effect in inverted type-ii semiconductors. Phys. Rev. Lett. **100**, 236601 (2008)
23. N. Marzari, A.A. Mostofi, J.R. Yates, I. Souza, D. Vanderbilt, Maximally localized wannier functions: Theory and applications. Rev. Mod. Phys. **84**, 1419–1475 (2012)
24. X.-L. Qi, Y.-S. Wu, S.-C. Zhang, Topological quantization of the spin hall effect in two-dimensional paramagnetic semiconductors. Phys. Rev. B **74**, 085308 (2006)
25. X.-L. Qi, S.-C. Zhang, Topological insulators and superconductors. Rev. Mod. Phys. **83**, 1057–1110 (2011)
26. R. Resta, Berry Phase in Electronic Wavefunctions. Troisieme Cycle de la Physique en Suisse Romande (1996)
27. R. Resta, Macroscopic polarization from electronic wavefunctions. arXiv preprint cond-mat/9903216 (1999)
28. R. Resta, What makes an insulator different from a metal? arXiv preprint cond-mat/0003014 (2000)
29. S. Ryu, A.P. Schnyder, A. Furusaki, A.W.W. Ludwig, Topological insulators and superconductors: tenfold way and dimensional hierarchy. New J. Phys. **12**(6), 065010 (2010)
30. S.-Q. Shen, Topological insulators: Dirac equation in condensed matter. Springer Ser. Solid-State Sci. **174** (2012)
31. A.A. Soluyanov, D. Vanderbilt, Smooth gauge for topological insulators. Phys. Rev. B **85**, 115415 (2012)
32. J. Sólyom, *Fundamentals of the Physics of Solids: Volume III: Normal, Broken-Symmetry, and Correlated Systems*, vol. 3 (Springer Science & Business Media, Berlin, 2008)
33. D.J. Thouless, Quantization of particle transport. Phys. Rev. B **27**, 6083–6087 (1983)
34. G.E. Volovik, *The Universe in a Helium Droplet* (Oxford University Press, New York, 2009)
35. F. Wilczek, A. Zee, Appearance of gauge structure in simple dynamical systems. Phys. Rev. Lett. **52**, 2111–2114 (1984)
36. D. Xiao, M.-C. Chang, Q. Niu, Berry phase effects on electronic properties. Rev. Mod. Phys. **82**, 1959–2007 (2010)
37. R. Yu, X.L. Qi, A. Bernevig, Z. Fang, X. Dai, Equivalent expression of Z_2 topological invariant for band insulators using the non-abelian Berry connection. Phys. Rev. B **84**, 075119 (2011)
38. J. Zak, Berry's phase for energy bands in solids. Phys. Rev. Lett. **62**, 2747–2750 (1989)